Series on Technology
and Social Priorities

NATIONAL ACADEMY
OF ENGINEERING

Information Technologies and Social Transformation

Bruce R. Guile
Editor

NATIONAL ACADEMY PRESS
Washington, D.C. 1985

National Academy Press • 2101 Constitution Avenue, NW • Washington, DC 20418

The National Academy of Engineering is a private organization established in 1964. It shares in the responsibility given the National Academy of Sciences under a congressional charter granted in 1863 to advise the federal government on questions of science and technology. This collaboration is implemented through the National Research Council. The National Academy of Engineering recognizes distinguished engineers, sponsors engineering programs aimed at meeting national needs, and encourages education and research.

Funds for the National Academy of Engineering's Symposium Series on Technology and Social Priorities were provided by the Andrew W. Mellon Foundation, the Carnegie Corporation of New York, and the Academy Industry Program. The views expressed in this volume are those of the authors and are not presented as the views of the Mellon Foundation, the Carnegie Corporation, the Academy Industry Program, or the National Academy of Engineering.

Library of Congress Cataloging in Publication Data

Main entry under title:

Information technologies and social transformation.

(Series on technology and social priorities)
Papers presented at a symposium held in conjunction with the 1984 annual meeting of the National Academy of Engineering.
1. Electronic data processing—Social aspects—Congresses. I. National Academy of Engineering.
II. Series.
QA76.9.C66I52   1985   303.4'834   85-4830

ISBN 0-309-03529-5

Printed in the United States of America

SYMPOSIUM ADVISORY COMMITTEE

Chairman

JOSEPH V. CHARYK, Chairman and Chief Executive Officer, Communications Satellite Corporation

Members

SOLOMON J. BUCHSBAUM, Executive Vice-President, Customer Systems, AT&T Bell Laboratories

HARLAN CLEVELAND, Dean, Hubert H. Humphrey Institute of Public Affairs, University of Minnesota

N. BRUCE HANNAY, Vice-President, Research and Patents, Bell Laboratories (retired)

JAMES BRIAN QUINN, William and Josephine Buchanan Professor of Management, Amos Tuck School of Business, Dartmouth College

Staff

BRUCE R. GUILE, *National Academy of Engineering Fellow*

PENELOPE GIBBS, *Symposium Series Secretary*

Preface

Information technologies are perhaps the most aggressive technologies of the current age, generating progress, change, and turbulence in many branches of industry and in the lives of organizations and individuals. Microelectronic systems are shrinking in size and cost, growing in performance, expanding their range to the world level, and crossing cultural boundaries. As societies attain higher orders of information handling, existing social institutions are faced with the dual challenge of directing and accommodating social change driven by technology. With this in mind the National Academy of Engineering (NAE) held the second of its Symposia on Technology and Social Priorities, in conjunction with its 1984 Annual Meeting, titled Information Technologies and Social Transformation.

The symposium, held on October 4, 1984, brought scholars of technology and society together with technologists, social scientists, and representatives from the industrial, legal, and public sectors to discuss the interaction of information technology with social institutions. The topics addressed included a review of recent developments and likely futures in information technology, a comparison of information technology to historical developments in other technologies, and discussion of the interaction of information technology with businesses, homes, property rights in information, and various hierarchies of social organization. The six papers presented at the symposium, with comments by discussants asked to prepare remarks on the papers, make up this volume. It is, I think, an exceptionally interesting collection of insights about the role of technology in society in general and the likely impacts of information technology in particular.

The Council of the National Academy of Engineering voted to dedicate the symposium to the memory of George M. Low, who died on July 17, 1984. In his 27 years with the National Aeronautics and Space Administration (NASA), Mr. Low made major contributions to the Mercury, Gemini, and Apollo programs as an engineer and an administrator. He became deputy administrator of NASA in 1969 and served in that position until he became president of Rensselaer Polytechnic Institute in 1976. Mr. Low was elected to the National Academy of Engineering in 1970 and was the recipient of the Academy's Founders Award in 1978. At his death Mr. Low was a member of the Council of the NAE and chairman of the joint National Academy of Engineering–National Academy of Sciences–Institute of Medicine Committee on Science, Engineering, and Public Policy.

Bruce Guile, a fellow at the National Academy of Engineering, was, in large part, responsible for the symposium. Working with the advisory committee, he helped formulate the content of the symposium, worked with the speakers to ensure cohesion of the presentations, and served as an editor to produce this volume.

The editor acknowledges with gratitude the counsel and help of several Academy staff members. Jesse Ausubel, special assistant to the president of the National Academy of Engineering, offered helpful advice and answered questions throughout the process of holding the symposium and editing the papers. Penelope Gibbs, in addition to typing correspondence and the manuscript, handled the lists and labels that make a symposium happen. Dorothy Sawicki of the National Academy Press offered sound editorial advice and prepared the manuscript for publication.

ROBERT M. WHITE
President
National Academy of Engineering

Contents

Information Technologies and Social Transformation

Introduction

JOSEPH V. CHARYK

The papers in this volume reflect an attempt to balance two competing views on the way that information technologies and social institutions interact. In particular, in designing the symposium upon which this volume is based we tried to strike a balance between the optimism often generated by technological potentials and the pessimism that sometimes accompanies a cursory examination of how technology may alter the quality of life. If there is a single theme to this volume it is a considered evaluation of the mutual adaptation between information technology and social institutions. How do businesses, families, and the legal system accommodate new technologies and what will change as a result of their accommodation? To what extent are developments in information technologies driven by the desire to do things differently in homes or businesses? Will the flow of information across national borders—allowed by advancing information technologies—change the character of international relations between industrialized nations?

In the abstract, questions concerning the rate and direction of mutual adaptation between technology and society are of importance to the scholars of technological evolution. On a less abstract level the same questions may be of interest to individuals in two ways. First, in their capacity as representatives of institutions individuals seek to understand the direction and character of change in their institutions and in the world in which the institutions function. Second, in their capacity as participants in society individuals are inherently interested in the future they are likely to experience. It is our hope that the theme of accommodation between technology and social institutions—reflected both within individual papers and between competing arguments in different papers—will be of interest to those reading in either capacity.

In the first paper John Mayo reviews the recent history and probable futures of several information technologies including integrated circuits, computing technology, software, and photonics. Particularly interesting is the overview of likely futures he provides by working through some relatively simple calculations of the physical limits on existing technologies. Additionally, Mayo discusses the forces that drive change (both social considerations and the push of technological possibilities) and the gates (economic considerations and social predispositions, for example) that control the rate of change. These elements—the driving forces and the gates—set the stage for discussing the changing character and use of existing technologies and allow the elaboration of possible scenarios for future development of emerging technologies.

In the second paper Melvin Kranzberg uses historical analysis to understand the social and technological changes that may be brought about by changing information technologies. Kranzberg compares the classical Industrial Revolution to the potential information revolution in our own time. Kranzberg's analysis leads him to conclude that we are indeed facing a social revolution and that the revolution is driven not only by changing information technologies but by interaction between social institutions and a wide range of technological innovations. Information technologies will evolve in, and contribute to, a social revolution brought on by rapid changes in energy, materials, and industrial management technology and made up of changing social norms, economic conditions, and attitudes toward science and technology.

Additionally, Kranzberg offers an interesting interpretation of cultural lag, the slowness with which social institutions respond to changing technology. If nothing else, cultural lag is evidence that culture imposes its will—albeit unconsciously and in a somewhat disorganized way—on the development and use of new technologies. The important insight that accompanies this observation is that technology is a "quintessential human activity . . . it bears the contradictions—the 'goods' and 'bads'—to be found in all complex human activities."

Mayo and Kranzberg, taken together, lay the foundation for the volume. Mayo focused specifically on changes in the technology of information handling. His unit of analysis is a particular technology, and he is persuasive about the direction of development of specific information technologies. Mayo is a technological insider taking a very informed look out from the specifics of information technologies; the power of his arguments comes from comparing the development of existing information technologies to the likely development of emerging technologies, and from comparing the potentials of competing emerging technologies. Kranzberg, in contrast, chose as his analytical unit not the technology but rather the process of change in the organization of social institutions. As a result, his paper places changes in information technology in a broader perspective in terms of other technologies and social institutions. Kranzberg stands somewhat outside the specifics of information technology and looks broadly both at the place any particular technology can hold in a social transformation and at the role that information technology may hold in current social transformations.

Mayo and Kranzberg together identify the tremendous potentials of information technologies and place those potentials in perspective both historically and relative to other current technological changes. Each of the other four authors takes a particular slice of the issue of interest by discussing the interactions between information technologies and different specific social institutions.

The papers by Harlan Cleveland and Anne Branscomb are almost exactly paired to explicate a debate over the adaptability of important social institutions. Cleveland, in his paper on the twilight of hierarchies, takes a broad look at five hierarchies that have served as means of social organization, and considers how advancing information technologies may erode the forces that have historically held those hierarchies together. He examines hierarchies of power based on control, of influence based on secrecy, of class based on ownership, of privilege based on early access, and of politics based on geography. He sees dramatic potential for change in a wide range of human endeavor as a result of emerging information technologies.

Anne Branscomb focuses on a specific social institution—the U.S. legal system—and examines how that institution will adapt to changing information technologies, particularly the information-based property rights accorded to individuals. Branscomb's focus is explicitly on the manner in which the legal system adapts. She finds, in her analysis of recent case law and legislation relating to property rights in information, evidence that the system is changing in a way that is consistent with the existing structure. Branscomb seems convinced that the current legal system will adapt effectively to changing information technologies.

Cleveland and Branscomb appear to disagree, but, on close examination, one finds that the disagreement is not over process but over degree. Both recognize that social institutions will adapt; it is in the nature of institutions to change in response to changing conditions, and some institutions, like the legal system, are specifically designed to be adaptive. The disagreement comes in hypothesizing about whether the rules that have guided and facilitated change in the past will be adequate to accommodate this particular dramatic technological change. The question lies in whether the changed form of an institution will be recognizable. Will it function in the same manner and be capable of serving the same goals? Branscomb asserts, implicitly, that the legal system is resilient and effective. Cleveland argues that the adaptation required by this new technology will fundamentally alter the character of many of our institutions—including how the U.S. legal system vests individuals with information-based property rights. Their disagreement over degree of adaptation will be settled, not surprisingly, only by time and experience.

Walter Baer and Theodore Gordon, like Anne Branscomb, work close to the interaction between information technologies and a particular social system. Baer, in his paper on information technologies in the home, divides the activities in the home into four major types—working at home, doing chores at home, learning at home, and relaxing at home—and considers the impact of more powerful and pervasive information technology on each type of activity. In each of the four cases Baer is cautious about the effects of technological advance. Particularly persuasive is his hypothesis that both time budgets and money budgets play an important role in determining the use of information technologies in the home. The time savings—rather than the money savings—realized by providing home buying services or financial services are perhaps the most important force pulling these services into the home. Additionally, new services brought to the home through emerging information technologies, especially entertainment services, will compete with television, radio, neighborhood softball games, and casual family conversation for a relatively scarce minute of available leisure time.

Gordon, in his paper on information technologies in business, selects four examples of the way that information technologies may affect business and develops each example by explaining a likely path for development and then hypothesizing about the consequences. Particularly interesting are Gordon's discussions of the impact of programmable automation on employment and of the implications of computer simulation for training. Though Gordon sees substantial change in business operations due to information technologies, he expresses some doubts that the character of business will change. In the end business still "takes raw materials, adds value, and sells products."

This is fundamentally the same cautious note that Baer strikes by consistently reminding the reader that information technologies are more likely to follow the desires and predispositions of individuals than to lead them.

Though there are similarities between the likely experiences of businesses and homes in adapting to information technologies, there are also important differences. The most important difference in likely experience derives from the fact that businesses are organized primarily for profit while homes exist to satisfy a wide range of individual human needs. Businesses, since they are organized to serve less diverse goals, may be more adaptive to opportunities provided by technology than are individuals or family units. The potentials of information technologies would seem to lie more, for example, in reducing the cost of business operations than in improving the quality of companionship or child rearing in the home.

Like any treatment of a broad topic, this volume does *not* address a number of issues of great interest. In particular:

- How will emerging information technologies affect educational institutions and federal, state, and local governments?
- How will international political activities be influenced through the potential availability of secure voice and teleconferencing facilities and data bases— the advent of sophisticated "hot lines"?
- Direct mail, made possible by the potentials of inexpensive computing, has already revolutionized constituency contact and fund raising in industry associations, professional societies, charitable organizations, and political coalitions. How will continuing development of information technologies affect the activities of these organizations in the 1990s?
- How will the merger of previously regulated telecommunications entities and competitive unregulated information-processing firms evolve? What will determine who will own, operate, and control high-cost facilities serving multiple needs where duplication of facilities is economically or operationally unrealistic?
- How will critical decisions be made as to allocations and use of naturally limited facilities (frequencies and satellite orbital locations)?
- Will restrictions be applied to the kinds of services that various entities are permitted to provide and, if so, is it possible to enforce them in the totally digital environment that is rapidly emerging?
- What forces will guide the emergence of communications standards and where will that guidance lead? What role will the standards that emerge play in determining the direction for the research and development that will bring us the next generation of information technologies? What role will state utility commissions and local port authorities play in determining what service will be available?
- As information flow across international boundaries explodes and political

control becomes more and more impotent in the new information technology world, what international organizations or mechanisms will emerge or should be created to ensure international order and cooperation?

- Are there public interest or universal service considerations that should be elements of U.S. information policy, and how will they be defined and implemented?

The list goes on. Nonetheless, the six papers in this volume, with the comments offered by the discussant for each paper, provide an introduction—and a few steps beyond—to the manner in which information technologies are forcing, and being shaped by, transformations in a range of social institutions. Even more importantly, the papers in this volume may offer examples of a manner of thinking about technology and social change that readers can use to understand the technology-driven social transformation where they work and live. That, anyway, is our hope.

This introduction would be incomplete without acknowledging the invaluable role played by Bruce Guile in planning, organizing, and bringing together the contributions represented in this volume. His energy and initiative were crucial to the success of this program.

The Evolution of Information Technologies

JOHN S. MAYO

THE INTERACTION OF TECHNOLOGY AND SOCIETY

Humans were given capable and inquisitive minds, so they endlessly seek better ways of doing things. This drive, coupled with an innate curiosity and a strong drive to unlock the secrets of nature, has created a steady stream of technical innovations over the ages.

These innovative efforts have focused on the means for survival, comfort, and accumulation of wealth—with the hierarchy of needs extending from physical basics of existence to higher-level wants associated with self-actualization. A principal thrust of innovation today continues toward technological advances that enhance the productivity of labor and free humans of tasks done more economically by machines. An insatiable appetite for convenience, comfort, and entertainment products and services, as well as for means to overcome natural barriers like geography and travel time, creates a constant pull on technology. The pull is especially strong in areas relating to the quality of life, and there have been many technical innovations to meet that need. But the opportunities are far from exhausted.

Among society's newest demands on technology is for the means to handle the vast amount of information generated by modern life. This information explosion stems from sophisticated business practices, new residential services, substantially increased record keeping through extensive data bases, and the globalization of our advanced society.

The information technologies have evolved over many years to assist a growing portion of the work force devoted to the generation, processing, transmission, storage, retrieval, and general use of information. Bureaucracies generated during the major wars and the rapid growth of social services in recent decades have helped increase the number of information workers in the U.S. work force, producing a permanent change in our way of life. Stimulated by these and other spurts of rapid growth, the percentage of information workers in the U.S. work force has grown from about 10 percent in 1900 to about 30 percent in 1940 to about 50 percent in 1970. Since 1970 the fraction has held at roughly 50 percent, probably as a result of the new electronic information technologies that augment human efforts. The computer, along with telecommunications, is making today's information work force more efficient, much as the engine raised productivity during the industrial revolution. In both cases, society's thirst for technology to reduce labor was met in striking ways by a wide range of innovations of varying impacts.

This thirst for technology creates a steady pull on innovation. In addition, the technologies themselves provide a push. From the families of all technologically feasible innovations of all time has come an almost endless reservoir of potential technology. However, between society's pull and the push of technology are two powerful gates, as shown in Figure 1. The technology available to society at any particular time is only that which can flow past the technology gate, which is

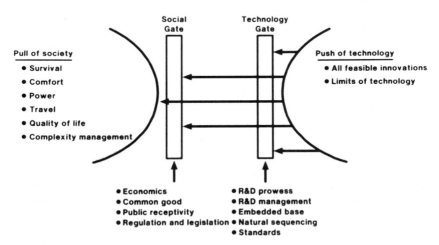

FIGURE 1   The flow of innovations into society.
SOURCE: AT&T Bell Laboratories.

operated by a series of strong forces. Among them are R&D prowess, characteristics of the embedded technology base, natural sequencing constraints, and perceived standards limitations:

- The force of *R&D prowess* is the sum of all the contributions of individual laboratories. The prowess of an R&D laboratory is limited by the skills of its scientists and engineers and by the capabilities of its support environment, including both financial and physical resources. Laboratory prowess also clearly depends on the wisdom and judgment of the R&D management team.
- The *embedded base of technology*, such as existing systems or production facilities that represent a large investment, impacts the characteristics of the R&D laboratories and the factories that make their products. It often leads the R&D laboratories to seek innovations that have great synergy with the embedded base, which can be either a curse or a blessing. On the one hand, this force can limit the introduction of new technology as well as discourage breakthroughs in totally new directions. On the other hand, it brings focus and resources. With good systems engineering, older technologies can be phased out, current ones upgraded, and entirely new ones introduced, all in ways that are synergistic with the embedded base.
- *Natural sequencing* simply means that the invention of the integrated circuit, for example, would have been unlikely before the invention and development of the transistor.
- The imposition of *standards* prior to innovation can narrow the technology gate by forcing R&D laboratories to focus on innovations that meet preconceived standards, but which may in the long run not be the best innovations at all.

Innovations that pass the technology gate must also pass the social gate. The forces that operate the social gate include economics, the common good, public receptivity, and regulation and legislation:

- The *economic force* depends not only on the marketplace, but also on the national economic structure. Currently, we see a significant difference in the way economic forces affect innovation in the United States as compared to Japan.
- Closely related to economics, but not always in concert with it, is the force that makes technology serve the *common good*. Society will eventually, for the most part, either ignore or legislate against technology that does not serve the common good.
- The issue of *public receptivity* is related to that of the common good. In the United States, the public defeated the supersonic transport and appears to have nuclear power on its deathbed; however, it still remains to be seen whether such innovations do not indeed serve the common good.
- *Regulation and legislation* have been and remain powerful forces at the social gate—forces very active in throttling technology on behalf of society.

Much social good has come from such actions, but not without frequent adverse impact in the long run.

The forces operating at the social gate are extremely powerful in selecting the innovations that actually succeed. They serve as a "tollgate" in the gap between the push of technology and the pull of society. The gating forces will be further examined, following the discussion of the information technologies pushing at the gate.

Technologies that survive both gates have primarily three types of impacts in the society they enter, depending heavily upon their character. First, of greatest impact are the "killer" technologies such as the engine, which replaced the horse, and the transistor, which replaced the vacuum tube. Their impact, of course, extends far beyond these major replacements, to opening whole new fields of opportunity. Included among these are opportunities to satisfy previously unknown or unrecognized societal needs and wants, often of an increasingly sophisticated nature.

Second in impact are the "new domain" technologies. Although they do not replace earlier technologies, they do open up entirely new areas of opportunity. An example of the new domain technologies is automatic speech recognition and synthesis, a rapidly developing technology that will eventually allow inanimate objects such as cars and appliances to speak and listen much as humans do.

Third, there are the "niche" technologies, which play a very important role in meeting society's needs. When they first appear, however, they are often mistaken for killer technologies. For example, when broadcast television became feasible, many expected it to kill newspapers, radio, and movies. Instead, it found its own niche and satisfied a thirst in society not previously met—or perhaps even recognized.

INFORMATION TECHNOLOGIES AND THEIR LIMITS

The growth in information jobs cited above is but one major indicator of the rapid transition of our society to an information base. Another major indicator is the rapid growth in information technologies. George Stibitz's invention of the first digital computer, the achievement of universal telephone service, and the invention of solid-state electronics paved the way to the Information Age. The technology of the Information Age is digital. The information is represented as digits, which are generated, processed, transported, stored, recovered, and displayed in order to do useful things for humans. The key technologies for manipulating digits are integrated circuits, computing technology, software, and photonics, as discussed below.

Integrated Circuits

The most powerful force of technology today is the expanding capabilities of silicon integrated circuits. A tiny chip of silicon can contain an electronic circuit consisting of hundreds of thousands of transistors and all the necessary interconnecting conductors—and its cost is only a few dollars. The circuitry on that chip is equivalent to about 10 years of work by a person soldering discrete components onto printed wiring boards. It is this tremendous improvement in the economics of circuit assembly, coupled with similar improvements in the reliability of individual circuit functions, that accounts for the power of this technology.

A common measure of progress in integrated circuit technology is the number of components that can be squeezed into a single-chip circuit. Figure 2 shows the exponential growth in components per chip over the past two decades, and a projection for the next decade. The number of components per chip of silicon is still increasing by a factor of 100 per decade. Today the limit is almost 1 million components on a chip; by 1990, it will be at least 5 million; and by the year 2000, between 10 and 100 million.

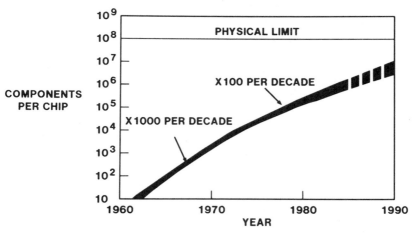

FIGURE 2   Changes in component density for silicon production, 1960 to 1990.
SOURCE: AT&T Bell Laboratories.

The limits of integrated circuit technology as we know it today are determined basically by how big we can make a chip and how small we can make the line widths used to define a working transistor. These limits can easily be estimated by assuming the largest practical chip to be about 1,000 square millimeters, and the smallest transistor to be fabricated using 0.1-micrometer (μm) line widths (a length of about 400 silicon atoms). When reasonable space for electrical isolation and interconnections is allowed, the resulting limit is easily derived to be about 100 million components per chip. For such a chip, using known technology, electrical isolation and interconnections would consume approximately 90 percent of the chip area.

The magic of the ever-expanding capabilities of integrated circuits will therefore be with us for at least another decade. Component reliability will continue to increase dramatically, and integrated circuit chips will perform more and more functions, ever faster and cheaper. This progress will make possible increasingly powerful, reliable, lower-cost digital systems and much more flexible approaches to systems design. Integrated circuit progress is making it possible to have digital systems everywhere, be they for computing, robotic control, office automation, or telecommunications. Clearly, this force will continue to be a major spur to further progress in the information technologies.

Computing Technology

Computing technology is a major beneficiary of the power of integrated circuits. Figure 3 shows the past trends in computer processing power and forecasts the future. Processing power is expressed in millions of instructions per second (MIPS), and each data point in the figure represents a specific computer introduced into the marketplace. Most notable is the rapid progress in processing power of the single-chip computer or microcomputer, which has been doubling each year. Extrapolation of the trend of the last decade forecasts that microcomputers will have processing power comparable to that of minicomputers and large, general-purpose mainframe computers (maxicomputers) by the early 1990s. Because the computer is the "engine" of the Information Age, having the power of today's largest computers on a chip or even a few chips of silicon to go in automobiles, appliances, toys, offices, factories, and homes is a tremendous driving force.

The trend curve of components per chip in Figure 2 showed that by 1990 integrated circuits will be within a factor of 10 of the physical limit of the technology known today. That implies the maxicomputers, minicomputers, and single-chip microcomputers as we now know them

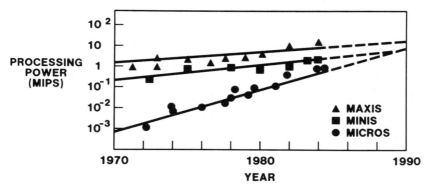

FIGURE 3 Trends in processing power of computers, 1970 to 1990.
SOURCE: AT&T Bell Laboratories.

have an ultimate limit of a few 10s of MIPS. However, the limit of chip processing power is not a limit to the processing power of computing. For computing is rapidly moving toward new architectures involving multiplicities of processing elements such as single-chip computers. Multiplicities of such chips, however, are inherently more expensive—largely because of the high costs and difficulty of interconnecting and programming them to function as single systems. For automobiles, appliances, tools, home computers, and the like, we can assume that most will operate with an ultimate computing power of the order of 10 MIPS or less per computer. But an automobile, for example, may eventually have a dozen or more computers.

Software

Software is vital not only to the operation of Information Age systems, but also to their interlinking with each other, with data bases, and with people. The demands for software are growing explosively—for tailoring systems to customers' needs, for making them reliable, and for making them easier to use, or "friendlier." These demands, in turn, are leading to increasingly complex software, ironically, to achieve user simplicity.

Unfortunately, software is the "bottleneck" information technology. Currently, it is generated principally by people, and most enterprises generate more software by hiring more people—a very difficult and costly approach. It still takes a programmer approximately one year to produce a few thousand lines of code. In telecommunications, a large electronic toll switching machine uses more than 2 million lines, a local electronic switcher more than 1 million lines, and System 85, a Private Branch Exchange (PBX), almost 2 million lines. Industry has learned in the last few years how to manage big software systems, developing them to meet cost, time, and performance objectives. But it desperately needs an improvement in programming productivity to sustain both the growth in complexity and the increasing demand for software systems.

The rate of progress in improving programming productivity remains extremely low. Figure 4 compares AT&T Bell Laboratories' productivity growth for producing software with that for design of silicon integrated circuit chips and circuit packs. The bad news is common throughout the industry. Software productivity is improving very slowly. The good news is the increasing productivity of the electronics designers, and that is largely a software success story. In fact, without computer-aided design, much of the progress in today's most important technology would be impossible—and tomorrow's tasks, hopeless. Even though chip complexity has increased 100-fold over the past decade, computer-automated tools permitted us to use the same design effort, as well as to significantly improve our ability to get error-free chip designs.

There will eventually be dramatic improvements in programming productivity. They will come from continued improvements in computer aids for software design, leading eventually to automatic gen-

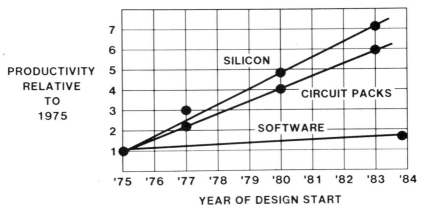

FIGURE 4   Hardware and software productivity, 1975 to 1984.
SOURCE: AT&T Bell Laboratories.

eration of applications programs. Such a breakthrough in software productivity will first require development of durable and detailed technical standards, new methodologies for requirements generation, and large software design programs for structuring, generating, and testing code. Progress may first come through greater development of reusable software components with standardized interfaces. Even those standards do not exist and will be difficult to achieve. Early circuit designers rapidly solved that problem for hardware components. Clearly, it can also be done for software components.

There is an argument, based on the analogy of motors, that today's software problems are transitory. When motors were new, users "hooked them up" to grinders, saws, and numerous other elements to create functional tools. But as that technology matured, users' needs were met by functioning systems that contained motors—drills, grinders, saws, washing machines, dishwashers, cars, toothbrushes, toys— an endless list. The analogy suggests that sooner or later a wide spectrum of software systems will be available so that most users will be able to buy functional Information Age products that perform the needed tasks. These products will just happen to contain software— much as dishwashers and refrigerators, for example, just happen to contain motors. The user could not care less, so long as the dishes are clean and the food cold. Such functional software-based products are rapidly emerging in the marketplace, but the trend has just begun.

Photonics

Photonics is the key Information Age technology for transmitting large amounts of digital information. There are two key innovations: the laser and ultrapure glass fiber. Combined, they provide a transmission capability that far exceeds that of copper wire and radio to meet the most stringent needs of the Information Age.

Photonics technology has progressed rapidly. In about a decade the technology has achieved some difficult technical milestones:

- Developing high-purity, ultratransparent, and high-strength glass fibers;
- Constructing long-life lasers that can operate at room temperature and at the appropriate wavelengths;
- Optimizing the mode of lightwave propagation in the fiber and shifting from multimode to single-mode fibers for many applications;
- Determining and exploiting the unique wavelengths at which fiber transmission losses are the lowest;
- Developing means for wavelength multiplexing of multiple bit streams onto the same fiber; and

• Developing single-frequency light sources which desensitize system performance to wavelength dispersion in the fiber.

Where is innovation in photonics leading lightwave systems? The current technical frontier is in increasing bit rates. The basic trend continues toward higher communications capacity per fiber and greater distances between signal amplifiers or repeaters. For example, since AT&T introduced the first full-service commercial lightwave system in 1977, fiber capacity has increased almost 10-fold, from 672 calls per fiber to 6,048. The corresponding amplifier spacings have increased from about 8 kilometers to more than 30 kilometers. In laboratory experiments described recently, AT&T Bell Laboratories set a "distance record" by transmitting 420 million bits per second over 125 miles without amplification. Also, 2 billion bits per second were transmitted over 80 miles using no amplification. That pulse rate can transmit the entire 30-volume *Encyclopaedia Britannica* in a few seconds. Underlying these accomplishments is healthy progress not only in glass fibers and lasers, but also in photodetectors and many other system components.

The limit of loss in today's fibers is shown in Figure 5. The two wavelength bands at which the loss of practical fibers is both low and near the limit are in the regions around 1.3 and 1.5 μm. Most new designs operate in these bands. Actual signal losses achieved in these bands are close enough to the theoretical losses that major new breakthroughs in silicon fiber performance do not seem likely. On the other hand, advances in materials processing could lead to entirely new materials systems for fibers. Also, the power output of lasers will rise, and the sensitivity of signal detection subsystems will improve. So getting from today's capability of 125 miles without amplifiers to a few thousand miles without amplifiers may yet be feasible. The difference is not spectacular for domestic communications. But the capability of sending signals a few thousand miles without amplifiers is significant in globalizing the Information Age, for it would enable us to span the oceans with passive lightwave systems.

What is the limit of lightwave technology as we know it today, and when will we reach that limit? Extrapolation of progress in the rate at which information can be sent through fibers and the distance it can travel without amplification, coupled with a little analysis, suggests the answers. Figure 6 shows that the product of rate in megabits per second (MBPS) and distance in kilometers (km) has been doubling yearly—and this will probably continue for the rest of the decade, at least. Each data point in the figure represents the leading edge

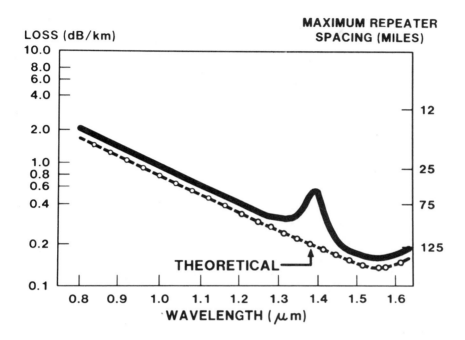

FIGURE 5 Lightwave communications technology.
SOURCE: AT&T Bell Laboratories.

accomplishment for a single wavelength channel—continually dominated by AT&T and the Japanese.

Simple detection theory can be used to estimate the physical limit
of today's lightwave technology. The estimate involves combining the
theoretical loss and nonlinear behavior of glass fiber with an assumed
maximum allowable laser power of approximately 1 watt and a minimum
requirement of about 10 to 100 photons per pulse. This forecast
suggests that the technology limits will permit the development of
lightwave systems with each channel operating 10 to 100 times faster
than today's best. Wavelength multiplexing will extend this limit by
another factor of 10 to 100, giving an ultimate limit of about 10^9, or 1
billion, MBPS/km.

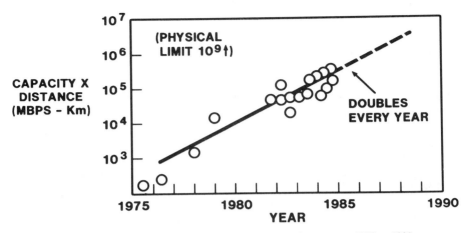

FIGURE 6   Lightwave technology capacity–distance achievements, 1975 to 1990.
SOURCE: AT&T Bell Laboratories.

The next technical frontier in reaching this limit is probably sophisticated multiplexing, modulation, and detection schemes. For example, quantized frequency modulation of the laser and coherent detection in the receiver may significantly expand the information carrying capacity of fiber systems.

TECHNOLOGIES LIKELY TO EMERGE

There is a vast reservoir of new technologies that will have significant impact on further evolution of the Information Age. They are discussed here in approximate rank order of importance. Among them is at least one potential killer technology. There are also areas where possible breakthroughs could extend today's technology into entirely new domains. And of more modest impact are new materials systems and clever innovations that create the niche technologies.

Potential Breakthroughs in Silicon Circuits

As noted earlier, the limits of today's silicon technology are basically determined by how small a working transistor can be made and by how large a chip can be made. The first possible breakthrough would be at the small end, where the current requirement is for a transistor defined by 0.1-μm line widths (a length of about 400 silicon atoms). Perhaps we will find a way to break this requirement and progress into the domain of molecular electronics—circuits based on ever-smaller devices, possibly down to individual molecules.

The second possible breakthrough would be at the large end, now limited by a chip 1,000 square millimeters in area. The ultimate breakthrough here would be defect-free wafers so that the chip itself could consume the whole wafer. Such wafer-scale integration would increase the number of components per chip by perhaps a factor of 100.

A third possible breakthrough would be a new, reliable, but cheap way to interconnect two or more integrated circuit chips so that they would operate as if they were one. This would enable a mosaic of simpler chips to do as much as or more than a single complex chip. Such a breakthrough would reduce the high costs of interchip connections, which is the force that now motivates us to squeeze increasing numbers of components on an individual chip. And it would open up totally new opportunities for radically different approaches to systems design and architecture. It would extend the magic of silicon technology beyond the edges of the monolithic chip, and it would fold the burden of system interconnection into the domain of photolithography and batch processing.

Integrated Optics

Integrated optics is a potential killer technology lurking at the gate. There is a possibility that most tasks currently done by electrons will someday be done by photons. So huge amounts of information may someday be transmitted, switched, and processed entirely by photons. Integrated optics systems are almost sure to emerge for at least specialized jobs. Before such all-photonic systems appear, however, combined electronic and photonic circuits will increasingly be integrated into the same circuit and possibly onto the same chip. Advantages include reduced parasitics, leading to higher speed, increased sensitivity, and optical interconnections. These result in reduced packaging limitations, less ground noise, and no induced noise in the optical elements. Such opto-electronic circuits also interface gracefully with lightwave transmission systems.

The next step, all-photonic integrated optics, is farther away. Much innovation will be required before optics can come close to matching the processing power of today's integrated circuits. Moreover, today's optical logic requires relatively high power, with the associated problems of heat dissipation. And optical logic is also physically large compared to integrated circuit logic. AT&T Bell Laboratories and others are working diligently to advance integrated optics technology. It may well become the same kind of driving force in the future as integrated circuits are today.

New Computing and Software Architectures

New architectures for both computing and software are surely feasible. Past thinking has been conditioned by the work of von Neumann and machine designs centered around the limited computing resources of one main processor and associated memory.

Vast new opportunities now exist for "nontraditional" computer systems architectures. Designers may use multiple processors, and they have the flexibility of distributing fast memory throughout the fabric of a machine's logic. Microprocessor chips will increasingly be integrated into complex distributed-processing architectures. This is a promising way to link a number of low-cost, relatively low-power microprocessors into a single system with the combined processing power of all the component chips.

Such a distributed-processing system, based on the limits of today's integrated circuit technology, places an enormous burden on software. For example, a good distributed-software operating system must coordinate all the separate processing units—perhaps as many as 1,000—as one system, yet permit each to work on a different problem. The software system must permit access to remote files or other data storage in a way that is transparent to the user, and it must permit remote execution of programs, independent of where the data reside and of which particular processing unit is working on the problem— again, transparent to the user. All this represents a very fine marriage between the capabilities of integrated circuits and the needs of distributed-software systems. The full potential of software architectures based on the limits of integrated circuit technology is yet to be fully understood. But it is an encouraging trend that distributed-software architectures are emerging at a fast rate. And these architectures increasingly look like the structures of the world's largest distributed computing network—the U.S. telephone network of thousands of interconnected switching machines.

User-Friendly Interfaces and Computer Speech

The full potential of the information technologies will not be realized until it is easier for people to use software-based systems—to deal with computers on their own terms and possibly in their own language.

It is interesting to note that even though the telephone network is the world's most complex computing system, the power of that vast network is available even to a child with virtually no training. Such powerful and simple ways to access the new machines of the Information Age do not yet exist.

A major breakthrough is near at hand. Machines that recognize and respond to human speech can now be built. Speech synthesis is well along, and speech recognition is progressing nicely. Vocabularies for recognition are currently limited, however, and some systems require "training sentences" from all users in order to recognize their speech. But the growing capability of integrated circuits is sure to produce eventually a few chips of silicon that are equivalent to portions of the brain, the ear, and the voice box. Then most machines and appliances will be able to speak and listen much as humans do. But speech recognition is a demanding course: whole sentences require 100 times more processing and about 10 times more memory than isolated words.

Data Networks

Achieving the full potential of the information technologies requires the interconnection of the physically separate computers made commonplace by silicon technology. Increased digital networking is required to interlink the computers not only with one another, but also with various large data bases. And this networking must include local clusters of homes and offices, as well as locations around the world. Fortunately, the telephone networks of the world are rapidly moving toward end-to-end digital connectivity. But they must be expanded to include more pervasive data packet switching. Importantly, networks must extend their software control to interconnect with the user's software, so that data network needs can be "dialed up" by the user and made immediately available—avoiding the delays of manual order processing and manual provisioning. There have been significant breakthroughs in such technology. The DACS, or Digital Access and Cross-connect System, first introduced some years ago by AT&T, is leading the way. Information Age telephone switching machines such as the AT&T 5ESS* electronic switch are adding even greater capability to meet the growing needs of automated data networks.

Integrated Circuits Based on Compound Materials

The integrated circuit industry is based mainly on silicon. A significant start has been made in a potentially competitive materials system,

---

*ESS is a trademark of AT&T Technologies, Inc.

gallium arsenide, which is but one of the many possible compound
semiconductor materials. Gallium arsenide has two features especially
worthy of note. First, electron speed is higher than in silicon, so
gallium arsenide devices of comparable dimensions are faster. Second,
because gallium arsenide is compatible with lightwave technology,
there is the potential of combining both electronic and photonic circuits
on a single chip. Figure 7 compares laboratory results in gallium
arsenide with the earlier production progress curve for silicon. The
rate of progress is impressive, but the leading edge of complexity of
gallium arsenide still has a long way to go to catch up with that of
silicon. Further, a saturation effect may set in, for as the feature sizes
of gallium arsenide integrated circuits become smaller and smaller, the
material's advantage of higher electron mobility diminishes with the
corresponding increase in electric field. It is simply too early to predict
the ultimate extent to which gallium arsenide will transfer significant
production away from silicon.

The potential of large-scale integrated circuits based on other
compound semiconductor materials is largely unknown. However,
AT&T Bell Laboratories recently reported the world's fastest semiconductor integrated circuit—which operates at 90 gigabits per second.
It is made of very pure gallium arsenide and aluminum gallium arsenide
in a multilayered, selectively doped heterostructure configuration.

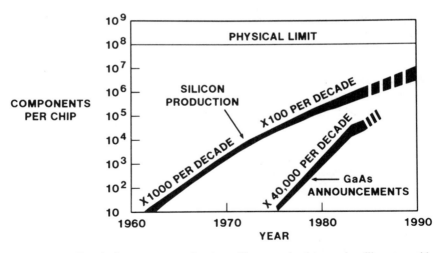

FIGURE 7 Trends in component density, silicon production, and gallium arsenide
announcements, 1960 to 1990.
SOURCE: AT&T Bell Laboratories.

Laser Materials Systems and Yields

The materials systems for lasers are in a relatively primitive state today, resulting in rather low manufacturing yields. Improved materials systems and structures would greatly lower the cost of lasers and, in turn, accelerate the pace of advancement of lightwave technologies. Because molecular beam epitaxy permits control of materials down to the atomic levels, it offers exciting and expanding opportunities to custom-fabricate new materials that may lead to new device concepts, including lasers.

TECHNOLOGY SELECTION AND THE PACE OF INNOVATION

Now that we have a view of the key technologies, their limits, and potential new technologies, let us further examine the gating forces that determine how the winning technologies will be selected and the resulting pace of innovation.

Marketplace Economics

The dominant force pulling innovations through the technology and social gates today is the needs of the marketplace. For decades electronic technologies have been pulled into the marketplace as fast as humanly possible. Today's technology, however, is so rich that it can do more things than society might find useful. Increasingly, marketing resources are required to sort out innovations and potential innovations, to contain the scope of development, and to focus investment on the applications that will win in the marketplace. This growing force of marketing in the information technology arena is creating new and challenging relationships. Such a give-and-take relationship between marketers and technologists has long operated in low-technology fields such as soaps and toothpaste. Today, there is a similar, rapidly evolving relationship in fields of highest technology, especially in computers, software, and telecommunications.

The pull of the marketplace rests largely on willingness to pay. To impact society significantly, an innovation must be of sufficient intrinsic value that users will not only pay the traditional manufacturing, sales, and related costs, but compensate for the high cost of development as well. Information Age products are very R&D-intensive. For example, R&D accounts for most of the cost of software, and the viability of software in the marketplace more strongly depends on sales volumes and copyright protection than do traditional manufactured products.

Technology selection is strongly tied to cost trends. For example, the proliferation of digital systems ties directly to the falling cost of a digital logic circuit, as shown by the curve in Figure 8. With each passing year, new digital systems become economically feasible because their costs drop below what the user is willing to pay. The figure also shows that at 1 dollar per logic gate, the telecommunications industry could economically justify two digital systems: one called T1, for voice transmission on wire pairs, and a digital controller for the 1ESS local electronic switch. At 10 cents per logic gate, the 4ESS electronic switch, a large digital toll machine, and the Dimension® Private Branch Exchange became feasible. At 1 cent per logic gate, 5ESS Information Age local digital switches became feasible, along with a wide variety of microprocessor-based "intelligent" telephones and terminals. Digital logic costs are now in the range of a tenth of a cent per gate and will be on the order of a millicent per gate by 1990. Each year as these logic costs continue to fall, the costs of a wide range of new Information Age products will drop below the threshold of user willingness to pay. The result is a mushrooming family of intelligent products—and the mushrooming phenomenon is not likely to slow down before the year 2000. The resulting economic climate will create a wide range of innovations, and economic forces will sort out the winners.

A user's willingness to pay is not an absolute. So technology selection is strongly affected by public opinion and advertising. The case of

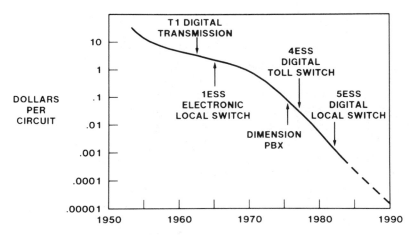

FIGURE 8   Digital integrated circuit costs, 1950 to 1990.
SOURCE: AT&T Bell Laboratories.

videotape recorders versus the videodisk is an interesting example, even though they are not functionally equivalent. It seems that society has selected the videotape machine as the winning technology, possibly because of its recording capability. The totality of forces that produced that selection is very difficult to characterize, but I suggest that the technologist needs help if he or she expects to forecast the technology selection process.

R&D Economics

The force of economics is somewhat stronger at the social gate than it is at the technology gate, but even so it is a powerful force behind our R&D laboratories and even our university research centers. For at their limits the key technologies are extremely expensive to develop.

The pace of innovation and, to a lesser extent, the direction of innovation are increasingly controlled by R&D economics. Clean rooms, feature patterning equipment, electron beam machines, and the like are terribly expensive at the micrometer level of capability and will become even more expensive as we move to submicrometer geometries. Small companies and universities are increasingly becoming followers rather than leaders. Today most universities cannot afford the requisite equipment, and even the wealthier ones are having to form special alliances with industry to raise the needed capital. The situation will ease somewhat as more universities find that complete fabrication facilities are not essential for strong teaching programs or even specialized research. But the trend is still there: the leading edge R&D that produces the significant innovations that guide all of industry can be afforded by only a few institutions in our society.

Government of course plays a large role in R&D economics. Of particular importance are the general economic climate of the nation, tax incentives, antitrust relief, and sharing of the output of the federal laboratories, as well as technology initiatives in the Department of Defense and the National Aeronautics and Space Administration. Each of these forces is managed on its own, but the sum of all the forces is not managed to speed the pace of technology and optimize the nation's technical position.

R&D Prowess

The prowess of our industrial, government, and university R&D laboratories remains a major factor in technology selection and the pace of innovation. Competence and motivation of individual scientists and engineers are vital to R&D prowess, but management of those laboratories also is very important. A prerequisite for innovation in a particular area is the dedication of people and capital resources to that area, and creation of an atmosphere conducive to innovation. AT&T Bell Laboratories is widely acknowledged for having the winning combination, as well as for creating most of the innovations for the Information Age. Their inventions—including solid-state devices, lasers, and a wide range of telecommunications and information technology—spawned new industries that have gained significant innovative strength of their own.

The power of these innovations to meet social needs is so great that they have forced change in some of society's major institutions. By opening up vast new frontiers of business opportunity and spawning numerous competitors, solid-state technology blended telecommunications and data processing and led to the restructuring of those industries, just as the engine forced restructuring in the industrial revolution. Industry structure, in turn, is a strong force in technology selection and in pacing innovation.

We are dealing with extremely powerful technologies whose forces are at best barely under our control. Continued U.S. industrial competitiveness is heavily dependent on the prowess of our R&D laboratories. Without a superior technical position, we could not be a force in technology selection or in the pace of innovation. The breakup of the Bell System has created a national challenge to ensure that the new environment has at least the R&D capabilities of the old. Certainly, AT&T is fully dedicated to continuing the innovative strengths of AT&T Bell Laboratories. However, the communications industry is much more fragmented than it was, and there are no means for managing it collectively. This is in sharp contrast to the major U.S. R&D competitor, Japan, which has numerous mechanisms for guiding its collective R&D efforts at the national level. Ironically, Japan has succeeded with its high technology by using methods, including statistical quality control, largely copied from the old Bell System.

Ensuring the future prowess of our R&D laboratories will take more than blind faith that competition among our R&D enterprises will spur results that are more innovative and cover wider domains than the collectively managed enterprises abroad. Success will come from meeting two broad challenges posed by the new environment. First, our industries in general must continue to devote the needed resources to ensure steady streams of innovation from each R&D laboratory. And second, in this new environment, each of the R&D laboratories must continue to add its innovations promptly to the advancement of science by continuing free and open dialogues within the technical communities.

Regulation

Few actions can throttle the flow of technology and slow the pace of innovation more than regulatory actions. In general, regulatory actions direct an entire industry or the most able within that industry. Actions are usually based on issues other than technology selection or pace, but the result may have enormous impact on technology. Regulating an entire industry always has the potential peril of handicapping every one of the affected domestic companies and of benefiting every one of the foreign competitors. Regulators must acknowledge that the nation's major innovations will continue to come from its most able enterprises, and that—through their action or inaction—they often have the power to throttle the outputs of these enterprises.

This throttling of technology comes from action as well as inaction. For example, legislative action ruled AT&T out of the international satellite business and thus led to the disbanding of the pioneering Telstar satellite development team, an act that hardly accelerated the flow of satellite technology to the marketplace. Regulatory inaction delayed the availability of cellular radiotelephones to the public by a decade after AT&T had the technology available.

At AT&T we are still experiencing a regulatory environment that restricts information flow within our business and favors competitors who have not contributed to the technology and have trivial R&D capabilities. Such regulation places us in an environment that is insensitive to the funding of R&D that can make a contribution not only to our business, but to the nation at large. As hard as we try to offset these handicaps, they remain a factor in our ability to maintain the flow of innovation.

Technical Standards

Technical standards are vital to the evolution of the Information Age. New products must not only be compatible with each other, but also with older products. And the standards must be firm enough to ensure compatibility, and prompt enough to ensure rapid introduction of new technology. Indeed, rapid progress seems to depend on prompt standards or no standards. Almost-at-hand standards encourage developers to wait and discourage researchers working on alternative options.

In the past, key industry leaders developed the technology and brought it promptly to the marketplace. AT&T technology rapidly became the standard for the telecommunications industry, and IBM set powerful standards in the computing field. In today's new environment, no single enterprise can pioneer the standards. For example, AT&T's UNIX℗ operating system grew out of telecommunications, but it is rapidly becoming a worldwide computing standard.

Furthermore, telecommunications has "shrunk" the globe, so standards must increasingly be global. In working to set global standards, the technologist has encountered severe political issues, such as "one nation, one vote." Attention to the global marketplace has created an increasing tendency to set up technical standards before new products emerge. That approach has great impact on technology selection and is sure to delay the flow of new technology because of the substantial risk in introducing products in parallel with the standards deliberations. Unfortunately, these frustrations, which have plagued the international standards scene for years, are now characterizing the domestic scene. The political motivations in domestic standard setting are greatly enhanced in the new environment, and the ability of domestic standards bodies to handle the issues is yet to be proven. A major challenge is to try to separate the political and technical issues, with the hope that the technical issues can be resolved promptly.

The Embedded Technology Base

Three examples illustrate the force of the embedded base: silicon circuits, lightwave systems, and magnetic bubbles.

The world's investment in silicon circuit fabrication facilities is estimated to be $15 billion. Most electronic products in the field are based on silicon, and the popular "direction" of the past decade has been, "If it can be done in silicon, it will." Until recently, almost 100 percent of the integrated circuit R&D was silicon based, partly due to the momentum of the embedded base, and partly due to the power of that technology.

Lightwave is a counterexample. Although there was no embedded base, enormous R&D resources have been devoted to that technology. The marketplace views lightwave as a killer technology, and tends to favor it over the embedded base.

Magnetic bubble technology was a potential killer that has been forced, at least for the present, to a relatively minor role. The power of the embedded base of magnetic disks and silicon random access memories drove cost reductions in these technologies as steep as the learning curve for bubbles. As a result, R&D in bubble technology diminished over the years and its learning curve slowed. So it will be a long time, if ever, before magnetic bubbles displace the embedded base. Josephson junction and other cryogenic devices appear to suffer the same fate, though they have never gotten as close to the competition as did magnetic bubbles. Both bubbles and cryogenics appear to be niche technologies at best.

THE INFORMATION AGE

Let me conclude by looking ahead. Considering the information technologies and their limits, pending new technologies, and forces controlling selection and pace, we can construct a pretty good picture of the technology of tomorrow. Whether this technology comes fast or slow, from our traditional R&D laboratories or from Japan or elsewhere, it is clearly leading us quickly into the Information Age.

What does this mean to society? This section considers some of the expected changes when information technologies will assist the mind much as the industrial revolution's technology assists the muscle.

Computers Everywhere

The industrial revolution produced the now mature age of motors— which we take for granted. All of us are surrounded by motors, from the large ones in our cars and boats to the small ones scattered about our households. As the Information Age matures, we will increasingly be surrounded by computers, from large ones handling major home or business tasks to a multitude of microcomputers in our cars, appliances, toys, games, entertainment centers—potentially everywhere. And we will think no more of computers lying idle in gadgets we hardly ever use, than we worry about unused motors in our homes today. Computers are becoming so widespread and low cost that we can afford to take them for granted, too.

Overcoming Geography

The Information Age has another dimension—bridging distance or geography. The interlinking of computers, people, machines, and data bases by the telecommunications network adds a new dimension of excitement. The Information Age will probably not offer electronic transport of matter, but it will do the next best thing. It will sense and transmit the remote environment to your presence via audio, video, and data. Such benefits depend in large part on the increasingly digital nature of the electronic world, with digital connectivity becoming universally available. Independent of geography, then, machines can talk to machines, machines can talk to people and people to machines, and—of course, as usual—people can talk to people, but with expanded options such as video conferencing.

It is increasingly feasible for people to conduct business from wherever they happen to be by accessing more and more capable telecommunications services and links back to their usual points of contact. One can even envision a "telepresence" using sensor-equipped robots to transmit perceptions back to the user. The user could "move" through scenes and even manipulate parts of the environment at long distance.

New Services

Information technology makes possible an expanding family of services such as financial transactions from home and office computers—and entertainment, shopping, and accessing of data bases by video and telephone links. Electronic mail and the ability to leave telephone messages are developing, though slower than the technology allows. Increasingly economical equipment can control energy use in auto, home, and office; and it can monitor security and report alarms by the automatic dialing of emergency services. A wide range of direct-dialing, teleconferencing, and announcement services are available on the telecommunications network.

The intelligent telecommunications network also makes it increasingly possible for users to have designated calls forwarded to another location, to have distinctive ringing identify calls from a specified set of telephone numbers, and to activate traces on nuisance calls. The network's intelligence also makes possible a wide variety of services such as the ability to dial one national number from anywhere in the country to be connected to the nearest emergency auto service, hospital, or any number of important services. Another potential innovation is to assign telephone numbers to people rather than telephones. By keeping the computer informed of the nearest telephone, you could designate calls to be routed automatically to you—wherever you might be, at your office, home, car, or at some hotel while on a trip. The same network intelligence is at the heart of cellular radio technology, which makes it potentially feasible for every car to have a telephone. This mobile phone capability is now spreading rapidly.

Of course it works just as well for personal portable phones as for car phones.

Universal Information Service

Early telephone pioneers wisely and quickly realized that achieving the full potential of the telephone depended on every phone being connectable with every other one. That realization is remarkable because it came when few people had phones, when phones were interconnected in small clusters to serve neighborhoods and businesses. And it came, despite the clear fact that any individual phone, over a lifetime of use, would actually need to connect to the tiniest fraction of the total telephone population.

So it is for individual computers. The concept of universal information service recognizes that every computer, data base, or smart terminal must be connectable with every other, even though most of the possible connections will never actually be needed. Today, clusters of data networks are growing in selected neighborhoods and many businesses. Fortunately, the telephone network is able to interconnect these clusters. But, for a variety of reasons, total connectivity falls far short of what is needed for universal information service. Achieving this goal will be a great challenge for the information technology and service industries—perhaps more so than universal telephone service was in the past.

Video Data Bases

High-speed digital transmission—using the ever-increasing information-carrying capacity of lightwave systems—could make video communications and pictorial data bases as widespread as today's telephone service. And this prodigious capacity could also be used to carry high-speed data for processing by intelligent machines. These machines could search huge quantities of transmitted data, selecting and storing only the particular information—perhaps a news item, stock quote, message, or even a movie—that fits their users' needs or interests.

Expert Systems

The combination of increasingly powerful integrated circuits, computing technology, and software should enhance knowledge and allow increasingly expert software to take over some tasks done by expert humans. An example is the expertise required to generate more accurate weather forecasts—ranging from global forecasting to the kind of area forecasting that would permit us to manage crops better and to accurately predict the landfall of hurricanes. These expanded technological capabilities will often be applied to create "expert" systems in services ranging from legal and financial advice to medical diagnosis—not necessarily to replace the doctor or lawyer, but to make such services available conveniently and inexpensively to large numbers of people. Expert systems exist today and perform specialized tasks in the telephone network.

Engineering Perspective

From an engineering perspective, the electronic equipment of the Information Age all looks much the same. It is digital systems made up mostly of integrated circuits mounted on printed wiring boards. The equipment is extremely compact in terms of number of gates per circuit board. Interface equipment is usually a keyboard, telephone lines, cathode ray tube, and/or liquid crystal display. Plasma panel and large liquid crystal displays will displace some of the cathode ray tubes, but not for full-motion video for some time. The equipment is increasingly reliable, and large systems will contain extensive diagnostic subsystems for maintenance. Most reconfigurations and rearrangements will be made via software rather than by manual operations. The equipment is increasingly lower in cost per function, so larger and larger systems will be of the throwaway type, much as low-function pocket calculators are today. The systems will increasingly be designed by machines through even more powerful computer aids to design. And, of course, the software content of the systems will grow, but hardware will also grow to ease software burdens. The challenge for engineers will shift toward the two ends of the spectrum of work: process development and computer-aided design systems to support design on the one end, and systems architecture and higher-level design on the other. The fast pace of technology, the exciting systems possibilities, and the expanding product opportunities put the Information Age designer in an enviable position.

CONCLUSION

The information technologies are far from exhausted, though the limits of the major technologies are well known. A number of new technologies are emerging, and progress should continue at a fast pace for at least a decade or two. With the possible exception of integrated optics, during this time evolution is not likely to be dominated by new killer technologies. More likely, the rapid pace of current developments will continue to create ever more favorable economics, and extend the known technologies into new domains. The forces that control the pace of innovation and technology selection are not likely to change substantially unless the restructuring of the telephone industry produces unexpected results or overseas competition forces government action. Good innovations will continue to be rapidly pulled into the market-place. The resulting richness of high-quality, low-cost technology should help create a better society—an Information Age with a host of new computing and telecommunications services to make life more pleasant, productive, and interesting.

Comments

ERNEST S. KUH
Professor of Electrical Engineering
University of California, Berkeley

I would like to begin by proposing a simpleminded model of technology evolution for the mathematically inclined. Using the state-space analogy, which is familiar to most young electrical, mechanical, and aerospace engineers, we may represent the interaction of the four key elements of technology evolution that John Mayo defines: (1) technology base, (2) research and development, (3) sequencing, and (4) standards.

In my proposed model, the state of the dynamic system corresponds to the technology base in Dr. Mayo's analysis; the input corresponds to R&D; the dynamics of the system correspond to sequencing; and finally, the set of constraints corresponds to standards. It might be possible then to use this analogy to introduce, for some technologies at least, a quantitative analysis of evolution through the technology gate. Models aside, the second part of Dr. Mayo's presentation gives a brief account of recent and prospective innovation in information processing technology. I would like to respond to that portion of the presentation with three comments.

First, that which impresses me the most are advances in lightwave tech-nology. When I worked at Bell Laboratories 30 years ago, I was designing repeaters for submarine cable using vacuum tube technology. The progress made during the last 30 years in transmission is remarkable.

Second, the technologies John Mayo did not discuss were such mundane things as the display technology, printers, and workstations. Though these technologies already play a major role in today's markets, I believe that their importance to scientific and engineering research and development—to the evolution of information technology—will be profound. The synergy between new technologies and research is immense. Just imagine that many engineers and graduate students will have at their desks the immense computation power, the convenience in communication of their ideas to co-workers, perhaps across the continent, and the luxury of observing three-dimensional color pictures to enhance their intuition. There is no question but that research and development will dramatically benefit from technological advance.

Third, Dr. Mayo did not say too much about universities. We all know the fundamental contributions made by the universities: the birth of computers, the start of artificial intelligence, the advances in very large scale integration (VLSI), and the excellent work in computer-aided design in microelectronics. Universities will continue to play a major role in the evolution of information technologies; the combination of experts—some of the best minds in the field—and young graduate students has proven to be a powerful force in basic research. As educators we have the responsibility to organize our institutions to preserve our strength in basic research and, in addition, to collaborate with industry for creative exploratory development.

It is interesting to note that in almost all areas Dr. Mayo touched upon, there exist physical limits, if research continues with the present mode of operation, strategy, or materials. That is why scientific research, and especially basic engineering research, is so crucial to us in order to make major breakthroughs. We at universities and in research centers should keep our goals high and far ahead in order not to fall into the trap of only working on problems of immediate application. While working closely with industry is crucial, we must maintain a balance, for our main aim is still to develop fundamental knowledge that will then lead to major breakthroughs.

In some areas it will be difficult for even major universities to keep ahead of industry because of the enormous cost of equipment and facilities. However, I believe there is a way out if we compare what we are facing now with physics research after the Second World War. Major research centers were created to fill this need. Engineering research centers, recently proposed by the National Science Foundation, are only a beginning. Certainly, many areas of high technology—the fifth generation of computers, the next phase of microelectronics research, and the flexible manufacturing system, for example—could benefit from major research centers associated with universities. It is up to us, in conjunction with the government and industry, to see to it that we have the research base we need.

John Mayo has laid the technological base for us by giving us a model for thinking about technological evolution and by providing an overview of information technologies and where they are headed. We must, of course, be interested in the implications of the Information Age on lives of nonprofessionals and workers in general, as well as in potential harms caused by the information technology, for example, the problem of invasion of privacy.

The Information Age: Evolution or Revolution?

MELVIN KRANZBERG

Every time we pick up a newspaper or a journal or listen to the news we learn about new technological developments heralding major sociotechnical changes: "Microelectronics Revolution," "Postindustrial Society," "Computer Revolution," "Automation Age," and so on. Since all of these involve the accumulation, manipulation, and retrieval of data by computerized electronic devices and their application to many facets of human life, it is no wonder that the headlines shout that computer developments are transforming industry and society to produce a new "Information Age."

Is this transformation evolutionary or revolutionary? After all, most technologies are evolutionary in the sense that they derive from prior developments. The steam engine did not emerge full-blown out of James Watt's brain, but was based upon Thomas Newcomen's engine, which in turn rested on still earlier attempts. Similarly, Gutenberg's invention of printing derived from a whole series of previous innovations—paper, block printing, inks, and movable type—which he put together in a new way. Indeed, virtually every major technological innovation can be shown to have been the outcome of evolutionary advance, in that historians can trace the elements comprising them far back in time.

Computers, the basis of the Information Age, find their origins in earlier devices, such as the ancient abacus, the seventeenth-century calculators of Pascal, the work of Charles Babbage in the nineteenth century, and Herman Hollerith's development of punched-card operations for the U.S. Census in the 1890s.[1]

Even though such technologies evolve over a long period of time, they can have revolutionary technical and social impacts even during the process of reaching full development and application. However, history indicates that changes in individual technologies do not by themselves have revolutionary sociocultural effects. Thus the medieval improvements in power sources—the introduction of the windmill and the waterwheel on a wide scale—did not produce a "revolution" because they remained based in a small-scale agrarian society. Most people continued to live in rural villages with farming as their chief occupation; hence there were no major changes in where and how people lived and worked.

Not until the eighteenth century did a whole series of technological innovations come together to produce the classical Industrial Revolution. Although popular opinion credits Watt's steam engine with starting industrialization, many of its elements, such as power-driven machinery, the factory organization of work, and specialization of labor, had already begun in the textile industry long before Watt.[2] Concomitant changes were occurring in mining and metallurgy, and transportation was being improved by the development of canals and roadways. Furthermore, the foundation of a national banking system and extension of joint-stock companies helped provide the capital and financial requirements for technical investment and commercial growth. The point is that a single major technical advance does not in itself constitute a technological revolution. There must be other and related technical advances plus major changes occurring in the political-economic-social-cultural context of the times.

Nevertheless, scholars delight in labeling an era by its most advanced technology, even when that technology is at first very limited in its application. For example, even though the "Age of Steam" is said to have begun with James Watt, for almost a century after Watt's engine more aggregate power was generated in Britain by waterpower than by steam; and it took nearly 100 years after Fulton's creation of the "Steamboat Era" before sailing vessels disappeared from oceanic commerce.

Similarly, the Wright Brothers at the beginning of this century began the "Era of Flight," but then it was postponed for another 25 years until Lindbergh's famous solo flight from New York to Paris; yet the "Aviation Age" really did not take off until after World War II. In similar fashion, the "Space Age" was said to have dawned with Sputnik, but more than a quarter of a century has elapsed since then, and we have scarcely begun to exploit space. That is indeed a long day's dawning!

Obviously, a single technological feat, no matter how much attention is showered upon it, does not by itself constitute a complete technological transformation. Indeed, one of the characteristics of a true technological revolution is that a great many innovations take place at about the same time. Their coming together creates a synergistic, indeed, explosive, impact upon the production of goods and services.

But technology does not occur in a vacuum. Instead, it takes place in a social matrix and interacts with society. Thus, despite the evolutionary nature of its individual technical components, the British Industrial Revolution marked a truly revolutionary transformation of society because it changed where and how people worked, lived, thought, played, and prayed.

For millennia, agriculture had been the chief source of production. The home-and-hearth was the center of work, education, social relationships, recreation, and, indeed, all life. The Industrial Revolution changed all that.

With the Industrial Revolution the factory became the workplace, and the city became the dwelling place. Family relationships changed as the father left home each day to earn wages in a factory while the mother stayed home with the children; other new social patterns emerged in the crowded cities, while some traditional institutions, such as the church, saw their hold on people's lives weakened in the urban environment. Technological and societal changes interacted, overturning old patterns of living, thinking, and working, and creating new institutional systems and cultural values.

Using the classical Industrial Revolution of the eighteenth and nineteenth centuries as our criterion, we learn that an industrial revolution consists of two chief elements: (1) a series of fundamental technical changes in the production and distribution of goods accompanied by—sometimes caused by, sometimes reflecting, but in any event, interconnected with—(2) a series of social and cultural changes of the first magnitude. Both elements must be present; a series of technological changes alone would not constitute an industrial revolution, nor would sociocultural changes without concomitant technological developments produce a new industrial era.[3]

To see if the much-heralded, incoming Information Age is truly a revolutionary phenomenon, let us analyze both the technological and sociocultural changes in the classical Industrial Revolution and see if parallel transformations are occurring today.

THE CLASSICAL INDUSTRIAL REVOLUTION

Looking at the main technical features of the classical Industrial Revolution, we find:

- the use of new basic materials, chiefly iron and steel;
- new energy sources, deriving from new prime movers and fuels, such as coal and the steam engine, and, later, electricity, petroleum, and the internal-combustion engine;
- mechanical inventions, such as the spinning jenny, the power loom, and machine tools, which increased production with a smaller expenditure of human energy;
- the centralized organization of work in the factory system, which entailed the further division of labor and specialization of function, and these, together with improved machines, making possible interchangeable parts and mass production;[4]
- the quickening of transportation and communication through the steamship, the steam locomotive, the automobile, and eventually the airplane; and in communications, the telegraph, telephone, and radio; and
- the development of a science of technology.[5]

In the nonindustrial technological sphere, agricultural improvements embodying many of the same technical changes made possible the provision of food for a larger population. All these technological developments involved larger use of natural resources, increased efficiency, and the low-cost, mass production and distribution of food, manufactured goods, and accompanying services.[6]

Not so incidentally, all these technical advances also involved information. After all, technology is a form of knowledge—knowledge of how to make and do things—which is why we sometimes refer to it as "know-how." Technology implies hands and minds working together to produce more efficient machines, processes, products, and services. All of these require the application of new and better information or at least the bringing together of old items of information in a new and different way. Thus, the industrial transformation of the eighteenth and nineteenth centuries was based upon the application of new and better information to improve traditional methods and machines and, in the process, to create new products and services. And their synergistic interaction accelerated the pace of change.

While political revolutions occur rather quickly—or at least can sometimes be assigned definite dates—sociocultural revolutions, involving deep-seated changes in the ways in which people work, think, and live, require somewhat more time for their effects to manifest themselves. Nevertheless, they too are revolutionary in their impact.

We can see that in the nontechnical elements—the economic-social-

political-cultural transformations—that accompanied and became part
of the classical Industrial Revolution:

- the decline of land as the chief source of wealth in the face of the immense
  wealth created by industrial production;
- political changes reflecting this shift in economic power, as well as new
  state policies corresponding to the needs of an industrialized, rather than
  agrarian, society; and
- sweeping demographic and social changes, including the growth of cities,
  the development of working-class movements (indeed, the birth of a whole
  new social class, the urban factory proletariat), and the emergence of new
  patterns of authority within the family and at work.[7]

There were other broad cultural transformations. Workers were
forced to acquire new and distinctive skills, and their relation to their
work shifted; instead of being craftsmen working with hand tools,
workers became machine operators, subject to factory discipline. Also,
there were major psychological changes in people's confidence in their
power over nature, and, of course, in hedonistic satisfaction. For
industrialization made possible a torrent of material goods, which
ultimately brought about a higher standard of living. Advances in
agriculture, combined with progress in medical knowledge and public
health measures, meant that hunger began to disappear as a major
threat in the industrially advanced nations. People lived longer—and
better, in terms of material goods.

This was indeed a revolution, because it transformed individual lives
and society. And it was an Industrial Revolution because the development of industrial technology provided the basis for the sociocultural
changes.

A CURRENT TECHNOLOGICAL REVOLUTION?

Are the technological and the sociocultural changes occurring in
relation to today's advances in computers of sufficient magnitude to
hail ours as a revolutionary "Information Age?"

Certainly the technical foundation has been built, including a change
in basic materials.[8] Let us remember that the introduction of new
technologies does not always mean the complete demise of older
technologies, especially in the case of materials. After all, wood
continued to be a major material source even when the Age of Steel
developed. While today's improvements in materials—composites,
plastics, synthetic fibers, sophisticated ceramics, and the introduction
of new alloys and lighter metals—do not mean that iron and steel are
outmoded any more than the coming of the Age of Steel meant that

wood ceased being used, these do represent a transformation in and an augmentation of materials resources affecting many other technical changes. Furthermore, the development of these new materials is roughly concomitant with the emergence of computer-aided design and manufacture. There is a synergy between technological developments as new materials find use in improving the operating effectiveness of the computers used to control manufacturing of the materials and manufacturing processes that work with the new materials.

In terms of energy, with the exception of hydroelectricity, the nineteenth century brought almost total reliance on fossil fuels. Within our own times, the fear that finite fossil fuels will eventually be exhausted has been somewhat alleviated by the possibility of almost limitless energy through exploitation of the power within the atom— although certain problems remain associated therewith. Also, greater emphasis is being placed upon conservation, synthetic fuels, renewable sources of energy, and greater and more efficient use of solar power. So although recurrent "energy crises" might come about through political and economic forces, we possess the requisite technical knowledge and potential to produce an abundance of energy in different forms. This represents a truly revolutionary technological advance over the fossil fuel era.

However, current changes in production mechanisms follow a somewhat different, yet nevertheless revolutionary, pattern than those of the past. The Industrial Revolution introduced power machinery and centralized production by multitudes of factory workers, and the early twentieth century further rationalized this process with Henry Ford's moving assembly line and Frederick W. Taylor's Scientific Management. But nowadays, computerized information devices form the heart—rather, the eyes, hands, and mind—of the machine and allow for completely automated machinery, robots. Instead of a machine operator, the human worker becomes a machine supervisor, overseeing a multitude of dials while the robotized machine—the steel-collar worker—does the actual work and replaces many blue-collar workers. Robots can perform dangerous operations, relieving humans from tasks that pose a threat to health and safety. They can also perform the monotonous and routine tasks which, some people claim, had made factory workers into machines.[9] The older mechanical devices had taken the burden off man's back; computerized devices also take the burden off man's mind.

In transportation too, information devices play a major role. So-phisticated jet engines—highly dependent upon electronic control and monitoring—have enabled airplanes to grow larger and speedier, replacing long-haul railroad and steamship passenger transportation. Also, we have completed the first voyages of exploration and are beginning to utilize space in new ways. These aerospace developments are linked with the microminiaturization of computerized information devices and are, indeed, dependent upon them. Still another example of the ubiquity of these revolutionary information devices is their application to the workings of automobiles and trucks performing very earthy tasks.

Communications too are being transformed, with satellite transmission of instantaneous information from all parts of the world. But that is only the most spectacular demonstration of how communication expertise has increased apace. Indeed, revolutionary advances in the flow, storage, manipulation, and retrieval of information, resulting from the improvements in computers, rightly entitle the future to be known as the Information Age.

These contemporary major technical changes—in materials, fuels and prime movers, machinery, the organization of work, transportation, and communication—all involve more knowledge and more information. Our industrial and agricultural technologies are increasingly reliant upon the newfound and enlarged technical capacity given us by computerized information devices.

As long as computers relied on vacuum tubes and were bulky, balky, and expensive, they had only a minor impact on industrial processes and structure. However, with the invention of transistors and their refinement into today's microchips, computers became omnipresent; their power was greatly multiplied, and they found many applications beyond computational number-crunching. It is this application of computerized information to all facets of life and technology that makes it the centerpiece of the new technological revolution.[10]

The computer has repercussions far beyond the field of information and computer science narrowly conceived. Civil, mechanical, textile, metallurgical, chemical, ceramic, and, of course, electrical engineering also make full use of our new informational capacity and expertise. The old slide rule hanging from the belt of the engineering student has given way to the pocket computer. Increasingly at every engineering institution in the country, the students have access to desk computers wired into larger computer systems. Indeed, computer literacy is no longer a monopoly of a small group of technical experts; instead it is being taught at the elementary school level, and it is fast becoming a necessary adjunct to liberal arts education, with personal computers becoming a ubiquitous item in educated households.

Just as the old Industrial Revolution transformed agriculture as well as industry, so today there have been revolutionary improvements in agricultural production. Less than 3 percent of the American population now lives on farms, and one American farm worker now produces enough food to feed 84 people. This is because agriculture itself has become thoroughly industrialized in methods and scale of production; like industry, it is being computerized in the breeding and feeding of livestock and poultry and in the growing of crops. Furthermore, the development of genetic technology to improve varieties of vegetables, fruit, and grain, to say nothing of livestock, rests upon biotechnological advances,[11] which in turn rely upon enhanced computer capabilities, as do new chemical fertilizers and pesticides. Agricultural technology is thus one of the chief beneficiaries of and contributors to the new Information Age.

The R&D laboratory, which grew out of the German chemical industry in the latter part of the nineteenth century, helped create a science of technology—or engineering science—and that is reflected in the education and practices of today's engineers.[12] Research and development, which has become characteristic of all technologically advanced industry, has, of course, been enhanced by our heightened informational capabilities. As a result our scientific/technical knowledge increases apace.

In brief, the Information Age has indeed revolutionized the technical elements of industrial society. But does it have similar revolutionary implications for nontechnical institutions, values, and society as a whole?

A CURRENT SOCIETAL REVOLUTION?

Let us look at some of the nontechnical changes that are occurring, partly as a result of the technological changes but also causing the advance of technology because of the synergistic relationship between technology and society. We can see that revolutionary changes are occurring in the pattern of industrial society, just as it marked a vast transformation from the preceding agrarian society.

Certainly, formidable economic changes are taking place which depart greatly from nineteenth-century industrial concentration. Although financial concentration is now occurring on an unprecedented scale, the economics and production technology of the older Industrial Revolution, which favored the consolidation of production, are now giving way to decentralized facilities—and on an international scale.

Henry Ford's River Rouge plant represented the peak of the older development: raw materials went in one end, and finished automobiles came out the other end. It was a marvel for its time, and people came from all over the world to see the wonders of "Fordismus." But no one ever built another River Rouge; instead, it was discovered to be more efficient and economical to disperse production facilities. Today's greater reliance upon more sophisticated materials and technologies reinforces the tendency toward dispersion—with, of course, profound impact upon the former centers of America's smokestack industries.

Similarly, when the first electronic computers were introduced some decades ago, their complexity, size, and expense seemed to dictate that the computerized information would perforce be concentrated and hence be susceptible to control by relatively few individuals. Indeed, this appeared to lend substance to George Orwell's vision of *1984* when all information—and hence all thought—would be controlled by "Big Brother." However, the introduction of the transistor and the development of the microchip allowed for the miniaturization of computing devices, so that today's small, hand-held computer can rival the past giants in information capacity and activity. As the young hackers at CalTech showed when they took over control of the scoreboard at the 1983 Rose Bowl game, the problem is no longer that Big Brother is watching you, but that "Little Brother" is messing up his program.

As a result, while the dispersion of information capabilities makes impossible the centralized control of information and the power implied therein, new problems regarding the secrecy of data, the patentability of software, and a whole host of new socio-legal problems confront us. We are still engaged in the process of discovering these new problems, and seeing if the old legal maxims still apply or whether we must work out new legal mechanisms to ensure a proper balance between private rights and the needs of the public.

Just as microcomputers make possible the diffusion as well as the centralization of information control, so industrialization, which had begun first on a regional, then on a national basis, is today being internationalized. Advancing technologies have made feasible the creation of new production centers, having different resource advantages, throughout the world. Partly this is due to the geographical dispersion of natural resources; today's sophisticated technology frequently requires exotic materials not available in the United States, so that we are no longer a self-sufficient nation producing all we need for our own uses and exporting to others. We even find it practical to import relatively commonplace energy supplies such as oil. Another resource advantage is lower labor costs, especially since some advanced manufacturing techniques, including those of assembling electronic devices themselves, oftentimes require only low skill levels on the part of production-line workers. The result is an internationalization of production of revolutionary dimensions, the implications of which are still not clearly discerned. However, it has led to a debate on "industrial policy" dealing with new mechanisms in order to provide training and gainful employment to those thrown out of work by automated manufacturing processes or by the transfer of production abroad.[13]

Yet, while employment in traditional industries declines, the statistics on the total number of employed people in the United States continue to mount. For, while computerized production technology allows us to produce a cascade of material goods with fewer workers, there has been an enlargement of the service sector of the economy. As a result, for the past 30 years more people have been employed in the service trades than in factory production, and the service sector continues to grow.

One reason is the enlargement of administrative and clerical activities, many of which derive from the heightened productive capability offered by automated devices and the consequent enhancement of service activities. Information automation in the office is proceeding apace,[14] and we historians, while having 20/20 hindsight, do not possess 20/20 foresight about its social impact.

Other writers, however, apparently possess a clearer vision of the future. For example, Alvin Toffler points out that computers will enable information workers to do their work at home, being tied in with central computers at the office.[15] Yes, it is indeed possible for more people to work at home. But the fact is that, with very few exceptions in certain occupations, such as editing and writing and the piece-rate processing of insurance forms and the like, that is simply not happening on a wide scale. The reason is that, as the ancient philosophers pointed out, man is a social and political animal. People like to congregate together; they derive intellectual stimulus and social satisfaction from personal contacts. The workplace is not only a spot for making a living but is also the site of the social interchange that is apparently a hallmark of our human species.[16] So, just because computers might offer us certain capabilities, this does not mean that we would want to take advantage of them, nor does it mean that they would necessarily be advantageous for the social interchange that, in the vast majority of cases, is essential for individual fulfillment.

Besides, Toffler neglects the fact that new technologies do not immediately and completely replace older forms. Instead, as we can see from the example of the classical Industrial Revolution, old technologies do not immediately die, nor do they quickly fade away. Instead, the new technologies are superimposed upon them and in many cases are used to augment the older capabilities.

My own guess is that we will be in the midst of the "Second-and-a-Half Wave" for a long time before we reach Toffler's "Third Wave," by which time the futurist scholars will already be talking about a "Fourth Wave."

Nevertheless, we can already foresee some possible changes in political and economic power. The old Industrial Revolution shifted political and economic power from the landed nobility, whose ownership of the land was the key to power and wealth in an almost totally agrarian society, to the industrialists. In England the new factory owners allied themselves with the old landed nobility to control the political apparatus. Yet at the same time the factory system, by concentrating workers, enabled them to organize and obtain considerable economic clout, not as individuals, but as a group. Then the enfranchisement of the workers in the industrially advanced states gave them a share in political power. In brief, industrialization carried with it political and social democratization—and the Information Age, by facilitating widespread communication, might conceivably fortify democratic political control in the advanced industrial nations.

Although we cannot be sure of that, we can be certain that governments will continue to be involved in economic policy and hence in technological activities. The nineteenth-century myth of laissez-faire blinded us to the fact that governments did in reality play a major role in developing the industrial economy: through tariffs to protect infant industries and by building or financing roads, bridges, and other elements of the transportation network and infrastructure. Indeed, the needs of a coordinated transportation system led not only to the adoption of a standard gauge for railroads but also to standard time zones. Furthermore, the increasing complexity of technology made governments encourage the development of measurement standards, such as for screw threads, and then safety standards. Today's sophisticated information technology has required further government action, often on an international scale, to assign radio frequencies and thereby allow for a freer flow of communications. In addition, the widespread use of more powerful chemicals and the fears of water and atmospheric pollution require governmental policing of safety standards in many industries.

Added to the technological need for governmental action is a growing public awareness of technology's importance to society, now and in the future, and hence the desire for some measure of public control.

Partly this is an outgrowth of a rising level of education, itself made possible through previous technological advance. As the Industrial Revolution began producing enough goods so that young children no longer had to be in the work force, they could be sent to school. Besides, the increasingly complex nature of technological devices required an educated work force.

As a result, we can trace the democratization of education throughout the nineteenth and twentieth centuries in the industrially advanced nations as a function of technological growth and complexity. At first elementary education became compulsory, then secondary education, and in the twentieth century America pledged itself to give equal access to higher education to all its citizens (sometimes irrespective of their ability to take advantage of it).

The new Information Age requires even more complex and sophisticated technology, so there is need for a still higher degree of specialized technical skills—including social skills as well as manipulative ones. Educational responses to the needs of the Information Age are already being discussed and fought over throughout the educational establishment—including, and perhaps especially, among engineering educators.

Still another revolutionary social change has been abetted by the new Information Age: the entrance of women into the work force in unparalleled numbers. Before the onset of industrialization, women worked alongside the menfolk in the fields and in the home handcraft production of the times. With the rise of the factory system and its regimen of disciplined work and hours, men became the breadwinners, while the women remained at home and were responsible for homemaking and child rearing.

However, machine technology has advanced to the point that brute strength is no longer a special asset, so women no longer labor under any physical disability. Machines do not know or care whether the hands that guide them are those of a man or a woman—or, for that matter, whether they are white, black, blue, purple, or green. As a result, advancing technology means that racial and gender distinctions scarcely matter in the actual production process—although, for social and cultural reasons such distinctions unfortunately persist in many parts of the world.

Women possess the physical stamina, intellectual qualities, and moral virtues that make them the equals of men in an Information Society where burdensome physical work has been taken over by machines. Hence, we are in the midst of a social revolution—some call it a sexual revolution—that is closely linked with the technical advances which have given women technical equality with men, even though they may not yet have acquired the social and political power that goes with their technical equality, to say nothing of wage equality.

Office automation will not only affect the clerical work that was the domain of women for almost the entire past century. Rather, it will extend to all aspects of production and distribution, since it allows for close monitoring of production processes as well as clerical tasks of billing and the like. Furthermore, it can give top managers fingertip access to information formerly supplied them by the middle managerial group. Here again, we cannot foretell with exactitude what will happen, but there will undoubtedly be further rationalization in the office procedures inherited from an earlier age, while the information user in the office will have more direct contact with the production process itself.

What is equally interesting to social historians and cultural anthropologists is that many of the revolutionary information devices will be incorporated into the mechanisms of our daily lives without our being aware of them. Already microchips are being used in the thermostats for our home heating and air conditioning systems and in the ignition and carburetion systems of our automobiles. But we will still set our thermostat at 70°, without awareness that the microchip is increasing the energy efficiency of our heating and air conditioning systems; and we will step on the gas or on the brakes without realizing that the microchip enables us to achieve better control of the automobile.

Of much greater significance than simply catering to our creature comforts are those major social changes occurring as an outgrowth of advancing information technology which will have a powerful effect upon our country's and the world's future. Among the most important are demographic changes resulting from public health, medical, and nutritional advances deriving from sophisticated computerized research in health technologies. As a result, people are living longer—and this is already changing the character of American society.

But there is a reverse side to this demographic coin, namely, rapidly exploding populations in the developing nations, where more than half the people are under 15 years of age. As a result, there are demands for technological development to meet the material needs of the world's growing population. At the same time there are apparently conflicting demands that this be done without plundering the earth of its resources or damaging the environment. In other words, the Information Age must stimulate technological growth to meet these demands and do so by new kinds of technical applications that will maintain the productivity and salubrity of our planet for future generations.

Finally, we come to the psychological changes, both social and individual, effected by technological changes. Until the Industrial Revolution people had always been fearful that the vagaries of nature would deprive them of life's necessities. With the plethora of material goods and foods made available through the technological advances of the nineteenth century, people were able to keep hunger at bay, and indeed overcome many of the hardships inflicted by nature through centralized heating and air conditioning systems, electrical lighting, and the like.

Not surprisingly, the world's fairs of the past century emphasized the great accomplishments of science and technology. The notion that human technical abilities would enable us to accomplish anything we attempted was given further credence some 15 years ago when man first set foot on the moon. Here was the culmination of the Scientific Revolution of the seventeenth century and the Industrial Revolution of the eighteenth and nineteenth centuries, the actual fulfillment of one of man's most ancient myths and dreams. It is no wonder that we could be accused of the old Greek sin of hubris, inordinate pride.

Paradoxically, however, at almost the very same time, we began discovering that many of our previous technological triumphs were despoiling the environment and that our military technology posed a threat to the continuation of life on our planet.

As a result, the new Information Age has brought with it a somewhat more equivocal view of the human relationship to nature. Instead of man's being the master of nature, it is now realized that man is a part of nature and that our future depends upon a fuller recognition of both nature's and humanity's capabilities and limitations.

But, that does not necessarily mean that doomsday is forthcoming, nor need it deprive us of hope. Unlike earlier ages when human technical capacities were prescribed by the availability of certain natural resources, limited in the forms of energy that might be applied, and constrained to do and to make things in the same way as their ancestors had done, our new technology provides us with many different ways of attacking problems. We now have many and growing options in regard to the materials that we wish to employ, the energy sources that we intend to utilize, and the ways in which we go about producing and distributing food, goods, and services. Because the scientific technology of the incoming Information Age offers us manifold choices, we can make decisions about the future course of society with due concern for conservation of natural resources, the preservation of the environment, and the well-being of our fellowman now and in the future.

TECHNOLOGY AND CULTURAL LAG

However, just because we have the ability to do new and wonderful things with our technology does not necessarily mean that we will actually do so. Many years ago the great sociologist William Fielding Ogburn postulated the concept of "cultural lag" in terms of human response to technical capabilities.[17] He pointed out that the technologies developed in the preceding century gave mankind the opportunity to bring about a new and better social system, allowing the vast quantity of material goods being turned out by an advancing technology to redound to the benefit to all of mankind, rather than being confined to a narrow few. However, he also stated that cultural systems and human institutions—governmental, legal, and the like—tend to lag in responding to new opportunities offered by these technical innovations.

Lewis Mumford's analysis, some 50 years ago, of the relations between technology and culture seemed to reinforce Ogburn's thesis.[18] He claimed that the latest technical innovations were still being employed to further the aims and goals of the earlier industrial transformation based upon the exploitation of nature and of human beings. In other words, while our technology might enable us to make a better world for all, it was being employed in the service of institutions and values belonging to an older and more selfish age, one that considered neither humanity nor the natural world.

The analyses of both Ogburn and Mumford were provocative when initially stated, but they appear simplistic in light of what actually happened. True, our new technology gives us capabilities to do many wonderful things, but we often continue to employ them in the service of institutions and values belonging to an older age. Mumford hoped our bright new technologies would point the way to a brave new world founded upon social justice and a concern for nature. Ogburn too felt that technology could better humanity's lot, and he deplored the "cultural lag" that prevented it from doing so. Both men implied that technology could do wonderful things for mankind, but things went wrong when we did not allow it to do so.

True, but what they forgot is that technology is a quintessential human activity, so it bears the contradictions—the "goods" and "bads"—to be found in all complex human activities. It is designed for human use, but that means it is also subject to human misuse and abuse. If technology were the sole determinant of human actions, our current world might be a much better—and certainly a different— place.

Here is an example of how an advance made possible by technology—

international goodwill through better communications and more contact among different peoples throughout the world—bogs down under the "cultural lag" afforded by nontechnical factors that take precedence over technical capabilities. Electronic messages can flow across the globe in a fraction of a second, irrespective of the political boundaries; hence the technical element of modern communication is indifferent to national boundaries. Similarly, there are no technical barriers to prevent airplanes from transcending national borders. In other words, modern communication and transportation have made nationalism technologically obsolete; however, any glance at the headlines convinces us that while nationalism might be technically obsolete, it still remains one of the most powerful forces affecting the future of mankind.

EVOLUTION AND REVOLUTION

Acknowledgment of this and similar facts has led me to reformulate the concepts of my predecessors who pioneered in analyzing the interactions between technical and sociocultural elements and has led me to formulate "Kranzberg's First Law." Kranzberg's First Law reads as follows: Technology is neither good nor bad, nor is it neutral.

By that I mean that technology's interactions with both the social and cultural milieus sometimes lead to developments that are far removed from the original goals of the technical elements themselves. For example, Henry Ford thought of his motorcar as a means to cheapen transportation and make personalized transport available to the masses. It did that of course, but it also did much more than that, transforming where and how we work, play, live, shop, eat, sleep, and—for those of you who remember rumble seats—even where we made love.

In accordance with Kranzberg's First Law, the Information Age will have similar and unanticipated impacts, as the computer goes far beyond the task of number-crunching and instantaneous communication of data. The variety of functions that computers serve suggests that their consequences will be mixed, unevenly distributed, and diffused, assimilated, and modified at uneven rates. Hence, we still cannot foresee exactly what some of the consequences will be, any more than the prophets at the turn of this century could foretell that the automobile would lead to the suburbanization of American society, provide the prototype for the mass production of all kinds of material goods, do away with the old distinction between city and country dweller, and, with its related industries, help produce the richest society in the world's history.

Furthermore, as a corollary to Kranzberg's First Law, the same technology can have quite different results when introduced into a different cultural setting. Thus, some technologies developed in advanced industrial countries have quite different effects when introduced into some developing nations. Because technology functions in a sociocultural matrix and depends upon an infrastructure that includes the educational level of the population, its political and economic institutions, and its value system (including religious beliefs), it can produce markedly different results when it interacts with a culture that differs from our Western industrial society.

The point I am trying to make is that this new Information Age presents mankind with many different possibilities. But because people differ historically in their cultural and social institutions throughout the world, the new technology can have quite different results when applied in differing sociocultural settings. Besides, the technology itself is still evolving, and hence might interact with our values, institutions, and attitudes along quite different lines than expected.

Even so, the historical record gives us some cause for optimism. The technical advances of the Information Age, if they follow the pattern of previous technical changes, could provide us with more goods and services, increase material well-being, and help do away with poverty and misery throughout the globe. And by giving us greater knowledge of the human, social, and environmental consequences of our technical options through the new informational tools available for technology assessment and impact analysis, the Information Age might help us avoid catastrophic assaults upon nature and upon our fellow human beings. For computer technology—along with its associated cluster of increasingly sophisticated analytic software, simulation models, and data bases—permits more complex analyses than have been previously possible in the social sciences. Indeed, the more information people have about nature, technology, and society, the more it might not only enable them to improve their living standards but also to do away with hatred and fanaticism—although we cannot be sure of that.

One thing we do know. Despite the many defects we can find in highly industrialized societies, including our own, the fact is that the most technologically advanced nations are the ones that have abandoned cruel and unusual punishments; have provided social welfare and medical services for all segments of society; have allowed for the greatest measure of racial, religious, and sexual equality; and have, in large measure, provided for freedom and a humane life for all.

The Information Age promises to carry those hopes for the good life even further. While it might be evolutionary, in the sense that all the changes and benefits will not appear overnight, it will be revolutionary in its effects upon our society.

NOTES

1. Although it was written before some recent, major developments, Jeremy Bernstein, *The Analytical Engine: Computers—Past, Present, and Future* (New York: Random House, 1964) provides a good popular account of computer history. See also Nancy Stern and Robert Stern, *Computers in Society* (Englewood Cliffs, N.J.: Prentice Hall, 1983). The *Annals of the History of Computing*, published by the American Federation of Information Processing Societies, contains articles about the recent as well as the "ancient history" of computers.
2. Terry S. Reynolds, "Medieval Roots of the Industrial Revolution," *Scientific American*, Vol. 251, No. 1 (July 1984):122-30.
3. Melvin Kranzberg, "Prerequisites for Industrialization," in Kranzberg and Carol W. Pursell, *Technology in Western Civilization*, 2 vols. (New York: Oxford University Press, 1967), Vol. 1, Chap. 13.
4. Although Britain was the birthplace of the Industrial Revolution, these developments were carried further in the "American System of Manufactures." See Otto Mayr and Robert C. Post, eds., *Yankee Enterprise: The Rise of the American System of Manufactures* (Washington, D.C.: Smithsonian Press, 1981); and David A. Hounshell, *From the American System to Mass Production, 1800-1932: The Development of Manufacturing Technology in the United States* (Baltimore: Johns Hopkins University Press, 1984).
5. A major article on this topic is Edwin T. Layton, "Mirror-Image Twins: The Communities of Science and Technology in 19th-Century America," *Technology and Culture*, Vol. 12 (Oct. 1971):562-80.
6. Standard accounts of the Industrial Revolution include David Landes, *The Unbound Promotheus: Technical Change and Industrial Development in Western Europe from 1750 to the Present* (London: Cambridge University Press, 1969); and T. S. Ashton, *The Industrial Revolution, 1760-1970* (Oxford: Oxford University Press, 1943).
7. See E. P. Thompson, *The Making of the English Working Class* (New York: Random House Pantheon Books, 1963); and Raymond Williams, *The Long Revolution* (New York: Columbia University Press, 1961).
8. Melvin Kranzberg and Cyril Stanley Smith, "Materials in History and Society," *Materials Science and Engineering*, Vol. 37, No. 1 (Jan. 1979):1-39; National Academy of Engineering, *Cutting Edge Technologies* (Washington, D.C.: National Academy Press, 1983), part III; Philip H. Abelson, "Materials Science and Engineering," *Science*, Vol. 225, No. 4675 (Nov. 9, 1984):613.
9. Larry Hirschhorn, *Beyond Modernization: Work and Technology in a Postindustrial Age* (Cambridge, Mass.: MIT Press, 1984).
10. See Tom Forester, ed., *The Microelectronics Revolution: The Complete Guide to the New Technology and Its Impact on Society* (Cambridge, Mass.: MIT Press, 1981).
11. Charles J. Arntzen, "Biotechnology and Agricultural Research for Crop Improvement," NAE, *Cutting Edge Technologies*, pp. 52-61.
12. Melvin Kranzberg, "The Wedding of Science and Technology: A Very Modern Marriage," in John Nicholas Burnett, ed., *Technology and Science: Important*

Distinctions for Liberal Arts Colleges (Davidson, N.C.: Davidson College, 1984), pp. 27-37.

13. A good summation of the issues involved is provided in Bruce Babbitt, "The States and the Reindustrialization of America," *Issues in Science and Technology*, Vol. 1, No. 1 (Fall 1984):84-93. Works featured in the debate include Lester C. Thurow, *The Zero-Sum Society: Distribution and the Possibilities for Economic Change* (New York: Basic Books, 1980); Bennett Harrison and Barry Bluestone, *The Deindustrialization of America: Plant Closings, Community Abandonment, and the Dismantling of Basic Industry* (New York: Basic Books, 1982); and Robert B. Reich, *The Next American Frontier* (New York: Times Books, 1983).

14. J. David Roessner et al., *Impact of Office Automation on Office Workers*, 4 vols., U.S. Department of Labor R&E Grant/Contract No. 21-13-82-13 (Atlanta: Georgia Tech Research Institute, 1983); Vincent E. Giuliano, "The Mechanization of Office Work," *Scientific American*, Vol. 247, No. 3 (Sept. 1982):148-64.

15. Alvin Toffler, *The Third Wave* (New York: Morrow, 1980). Similar optimism about the future role of information technology is to be found in John Diebold, *Making the Future Work: Unleashing Our Powers of Innovation for the Decades Ahead* (New York: Simon and Schuster, 1984).

16. Sherry Turkle, *The Second Self: Computers and the Human Spirit* (New York: Simon and Schuster, 1984) provides an interesting discussion of this point.

17. William Fielding Ogburn, *On Culture and Social Change: Selected Papers*, edited by Otis Dudley Duncan (Chicago: University of Chicago Press, 1964).

18. Lewis Mumford, *Technics and Civilization* (New York: Harcourt, Brace and World, 1934).

Comments

GUNNAR HAMBRAEUS
Chairman
Royal Swedish Academy of Engineering Sciences

It is my firm conviction that we are only at the beginning of a tremendous development which, in its effect on the individual and on society, will be more far-reaching than anything that we have witnessed until now.

The following three facts support my belief. First, we cannot yet discern any slackening of the pace in hardware development, as illustrated in Dr. Mayo's paper. This pace is in speed of operations, storage capacity, and reduction in price. Possibly we have not yet passed the point of inflection on the traditional growth curve.

Second, we still only utilize a small fraction of the capabilities of our hardware. The reason is, of course, the lag in software production and systems architecture. Ultimately software improvements will increase the productivity of present existing computers at least 10-fold. The combined effects of machine and program development will indeed be dramatic.

Third, the computer in combination with instant communications will multiply research and development productivity in all fields of science and technology. Already, data logging systems make possible the harvesting and interpretation of primary experimental data on a scale that we did not dare to dream of 20 years ago. In these vast collections of data, software can now trace connections and interdependences in a multidimensional space hitherto inaccessible to even the greatest giants of science. Hypotheses, theories, and conclusions can then be instantly conveyed to other research centers to be discussed, analyzed, and tested within days of the original new idea.

Also, computer modeling can, to a certain extent, substitute for real experiments, thus further increasing the speed of discovery. The same mechanisms also influence the application of knowledge in industry. Already we have a set of tools for development, design, production planning, and technology management that do for the engineers what numerical control and automation have already done for the worker on the factory floor. Here again, most of the opportunities are still before us.

The social consequences of this evolution will be far-reaching, to say the least. The important social consequences are not only in the various aspects of a cashless, paperless, and robotized future. There is the eternal struggle, political and social, over powerful tools. If you doubt my words, just witness the Soviets' frantic efforts to put their hands on information hardware. There are also important social implications for the rights of the individual vis-à-vis society, the questions of the integrity of the individual and the manipulation of people by misuse of information technology.

Information technologies carry both enormous benefits and grave dangers. We can have freedom or slavery, and a disturbing fact is that these options are poorly understood by laymen, the public, mass media, and decision makers. These issues are seldom the topic of serious discussion. I noticed with wonder the total lack of discussion on these issues in the 1984 U.S. presidential election race, and I find the same lack of interest in my own country. There is no more important task for an academy of engineering than to bring these issues into the open.

The Twilight of Hierarchy
Speculations on the Global Information Society*

THE INFORMATIZATION OF SOCIETY

It is still shocking, forty years later, to remember that the Manhattan Project, the huge secret organization that produced the atom bomb during World War II, did not employ on its staff a single person whose full-time assignment was to think hard about the policy implications of the project if it should succeed. Thus no one was working on nuclear arms control—though I. I. Rabi says he and Robert Oppenheimer used to discuss it earnestly over lunch. We have been playing catch-up, not too successfully, ever since.

The Manhattan Project was not an exception; it was the rule. For 300 years until the 1970s science and technology were quite generally regarded as having a life of their own, an inner logic, an autonomous sense of direction. Their self-justifying ethic was change and growth. But in the 1970s society started to take charge—not of scientific discovery but of its technological fallout. The decision not to build the SST or deploy an ABM system even though we knew how to make them, the dramatic change in national environmental policy, and the souring of the nuclear power industry bear witness.

The most prominent and pervasive consequence of the people's concern about the impacts and implications of new technologies is what the French call *l'informatisation de la société*. The made-up

*Copyright 1985 by Harlan Cleveland.

55

word, which we will Americanize to informatization, will serve as well as any to describe what is happening to some of our key concepts and conceptions as information becomes the dominant resource in postindustrial society. The new word is certainly better than postindustrial, which describes the future by saying it comes after the past.

The revolutions that began with Charles Babbage's analytical engine (less than 150 years ago) and Guglielmo Marconi's wireless telegraphy (not yet a century old) started on quite different tracks. But a quarter of a century ago, computers and telecommunications began to converge to produce a combined complexity, one interlocked industry that is transforming our personal lives, our national politics, and our international relations.

The industrial era was characterized by the influence of humankind over *things*, including nature as well as the artifacts of man. The information era features a sudden increase in humanity's power to think, and therefore to organize.

The information society does not replace, it overlaps, the growing, extracting, processing, manufacturing, recycling, distribution, and consumption of tangible things. Agriculture and industry continue to progress by doing more with less through better knowledge, leaving plenty of room for a knowledge economy that, in statistics now widely accepted, accounts for more than half of our work force, our national product, and our global reach.

A DOMINANT RESOURCE, A DIFFERENT RESOURCE

The size and scope of the information society are now familiar even in the popular literature. We can take it as read that information is the dominant resource in the United States, and coming to be so in other advanced or developed countries. To take only one cross section of this startling shift, the actual production, extraction, and growing of things now soak up a good deal less than a quarter of our human resources. Of all the rest, which used to be lumped together as services, more than two-thirds are information workers. By the end of the century, something like two-thirds of all work will be information work. Table 1 shows one effort at describing the sweep of change.[1]

It is not only in the United States that the informatization of society has proceeded so far so fast. A study by the Organization for Economic Cooperation and Development (the club of richer nations, with head-

TABLE 1   U.S. Work Force Distribution

	1880	1920	1955	1975	2000 (est.)
Agriculture & extractive	50%	28%	14%	4%	2%
Manufacturing, commerce,					
industry	36	53	37	29	22
Other services	12	10	20	17	10
Information, knowledge,					
education	2	9	29	50	66

SOURCE: © 1984 New Jersey Bell. Reprinted, with permission, from the Spring 1984 issue of *New Jersey Bell Journal.*

quarters in Paris) puts the average information labor force of several of its members' countries at more than one-third of the total during the early- to mid-1970s, and rising: the information component of labor increased its share of the total by 2.8 percent for each five-year period since World War II.[2]

Farming, which in some people's vocabularies is the most primitive of pursuits, is probably farther ahead than most industries in the embedding of information in physical processes. Says agricultural economist G. Edward Schuh: "All of the increase in agricultural output from the mid-1920s through the mid-1970s came about with no increase in the capital stock of physical resources. It was all due to new knowledge or information. That makes clear the extent to which knowledge is an output or resource."[3]

If information (organized data, refined into knowledge and combined into wisdom) is now our "crucial resource," as Peter Drucker describes it,[4] what does that portend for the future? Thinking about the inherent characteristics of information provides some clues to the vigorous rethinking that lies ahead for all of us:

Information is expandable. In 1972, the same year *The Limits to Growth* was published, John McHale came out with a book called *The Changing Information Environment*,[5] which argued that information expands as it is used. Whole industries have grown up to exploit this characteristic of information: scientific research, technology transfer, computer software (which already makes a contribution to the U.S. economy that is three times the contribution of computer hardware), and agencies for publishing, advertising, public relations, and government propaganda to spread the word (and thus to enhance the word's value).

The ultimate limits to growth of knowledge and wisdom are *time* (time available to human minds for reflecting, analyzing, and integrating the information that will be brought to life by being used) and the *capacity* of people—individually and in groups—to analyze and think integratively. There are obvious limits to the time each of us can devote to the production and refinement of knowledge and wisdom. But the capacity of humanity to integrate its collective experience through relevant individual thinking is certainly expandable—not without limits, to be sure, but within limits we cannot now measure or imagine.

Information is not resource-hungry. Compared to the processes of the steel-and-automobile economy, the production and distribution of information are remarkably sparing in their requirements for energy and other physical and biological resources.

Investments, price policies, and power relationships which assume that the more developed countries will gobble up disproportionate shares of real resources are overdue for wholesale revision.

Information is substitutable. It can and increasingly does replace capital, labor, and physical materials. Robotics and automation in factories and offices are displacing workers and thus requiring a transformation of the labor force. Any machine that can be accessed by computerized telecommunications does not have to be in your own inventory. And Dieter Altenpohl, an executive of Alusuisse, has calculations and charts to prove that, as he says, "The smarter the metal, the less it weighs."[6]

Information is transportable. In modern communication systems information travels at close to the speed of light. As a result, remoteness is now more choice than geography. You can sit in Auckland, New Zealand, and play the New York stock markets in real time—if you do not mind keeping slightly peculiar hours. And the same is true, without the big gap in time zones, of people in any rural hamlet in the United States. In the world of information-richness, you will be able to be remote if you want to, but you will have to work at it.

Information is diffusive. It tends to leak—and the more it leaks the more we have. It is not the inherent tendency of natural resources to leak: jewels may be stolen; a lump or two of coal may fall off the coal car on its way east from Montana; and there is an occasional spillage of oil in the ocean. But the leakage of information is wholesale, pervasive, and continuous. In the era of the institutionalized leak, monopolizing information is very nearly a contradiction in terms. Information monopolies will exist, as time passes, only in more and more specialized fields, for shorter and shorter periods of time.

Information is shareable. Shortly before his death, Colin Cherry wrote that information by nature cannot give rise to exchange transactions, only to sharing transactions.[7] Things are exchanged if I give you a flower or sell you my automobile. When our exchange transaction is done, you have it and I do not. If I sell you an idea, however, or give you a fact, we both have it. An information-rich environment is thus a sharing environment. That need not mean an environment without standards, rules, conventions, and ethical codes. It does mean that the standards, rules, conventions, and codes are going to be different from those created to manage the zero-sum bargains of market trading and traditional international relations.

THE EROSION OF HIERARCHIES

I am not a scholar of information/communication theory, but in my listening and reading as a practicing generalist I am struck with three seminal ideas as containing the most nourishment for our purpose, which is to think about how the new information environment is likely to modify our inherited assumptions about rule, power, and authority.

One is that information (in its generic sense) is not *like* other resources, nor, as some would have it, merely another form of energy. It is not subject to the laws of thermodynamics, and efforts to explain the new information environment by using metaphors from physics will just get in our way.

A second idea I find nourishing is that the ultimate purpose of all knowledge is to organize things or people, arrange them in ways that make them different from the way they were before. This is true of rearranging the genes in a chromosome, and it is equally true of rearranging people's ideas to create a movement. There is no such thing as useless knowledge, only people who have not yet learned how to use it. This was the powerful message carried in a 1979 article in *Science* by Lewis Branscomb, chief scientist of IBM. He wrote that information is so far from being scarce that it is in "chronic surplus." There is still plenty for scientists to find out, but "the yawning chasm is between what is already known by some but not yet put to use by others."[8]

A third insight, from the late British communications theorist Colin Cherry, is the distinction between the information ("message") itself and the service of delivering it. You may own the paper you hold in your hand, but you do not own its contents, the facts and ideas in the paper. Neither, now that I have written them down and you and I are sharing them, do I.[9]

The historically sudden dominance of the information resource has, it seems, produced a kind of theory crisis, a sudden sense of having run out of basic assumptions. This is only partially the product of information and communication technologies and their fusion in the new systems that are sprouting daily. Other dramatic extensions of scientific rationalism and engineering genius such as nuclear fission and gene splicing—all with an indispensable assist from the new information technologies—have also made their contribution to the *bouleversement* of long-held social and political convictions.

But somewhere near the center of the confusion is the trouble we make for ourselves by carrying over into our thinking about information (which is to say *symbols*) concepts developed for the management of *things*—concepts such as property, depletion, depreciation, monopoly, unfairnesses in distribution, geopolitics, the class struggle, and top-down leadership.

The assumptions we have inherited are not producing satisfactory growth with acceptable equity either in the capitalist West or in the socialist East. As Simon Nora and Alain Minc wrote in their landmark report to the president of France: "The liberal and Marxist approaches, contemporaries of the production-based society, are rendered questionable by its demise."[10]

The most troublesome concepts are those that were created to deal with the main problems presented by the management of things—problems such as their scarcity, their bulk, their limited substitutability for each other, the expense and trouble in transporting them, the paucity of information about them (which made them comparatively easy to hide), and the fact that, being tangible, they could be hoarded. It was in the nature of things that the few had access to resources and the many did not.

Thus, the inherent characteristics of physical resources (both natural and man-made) made possible the development of hierarchies of *power based on control* (of new weapons, of energy sources, of trade routes, of markets, and especially of knowledge), hierarchies of *influence based on secrecy*, hierarchies of *class based on ownership*, hierarchies of *privilege based on early access* to valuable resources, and hierarchies of *politics based on geography*.

Each of these five bases for discrimination and unfairness is crumbling today—because the old means of control are of dwindling efficacy; secrets are harder and harder to keep; and ownership, early arrival, and geography are of dwindling significance in getting access to the knowledge and wisdom which are the really valuable legal tender of our time.

Out of dozens of assumptions requiring a newly skeptical stare in the new knowledge environment, these five seem to me to bear most directly on leadership and management, because they are likely to affect most profoundly the ways in which, and the purposes for which, people will in future come together in organizations to make something different happen.

POWER BASED ON CONTROL: POWER AND PARTICIPATION

Knowledge is power, as Francis Bacon wrote in 1597. So the wider the spread of knowledge, the more power gets diffused. For the most part individuals, corporations, and governments do not have a choice about this; it is the ineluctable consequence of creating—through education—societies with millions of knowledgeable people.

We see the results all around us, and around the world. More and more work gets done by horizontal process—or it does not get done. More and more decisions are made with wider and wider consultation—or they do not stick. If the Census Bureau counted each year the number of committees per thousand population, we would have a rough quantitative measure of the bundle of changes called the information society. A revolution in the technology of organization—the twilight of hierarchy—is already well under way.

Once information could be spread fast and wide—rapidly collected and analyzed, instantly communicated, readily understood by millions—the power monopolies that closely held knowledge used to make possible were subject to accelerating erosion.

In the old days when only a few people were well educated and knowledgeable, leadership of the uninformed was likely to be organized in vertical structures of command and control. Leadership of the informed is different: it results in the necessary action only if exercised mainly by persuasion, bringing into consultation those who are going to have to do something to make the decision a decision. Where people are educated and are *not* treated this way, they either balk at the decisions made or have to be dragooned by organized misinformation backed by brute force. Recent examples of both results have been on display in Poland.

This is the rationale for Chester Barnard's durable theory of the executive function: that authority is *delegated upward*. As director of an organization, you have no power that is not granted to you by your subordinates. Eliciting their continuous (and, if possible, cheerful) cooperation is your main job as director; without it, you cannot get accomplished the most routine tasks (for which others are holding you, not your staff, responsible).[11] Indeed, nowadays in many offices orders that used to be routinely accepted are now resisted or refused.

In the modern American office, if you want a cup of coffee you do not take that co-worker, your secretary, off her or his own work to get it for you.

In an information-rich polity, the very definition of control changes. Very large numbers of people empowered by knowledge—coming together in parties, unions, factions, lobbies, interest groups, neighborhoods, families, and hundreds of other structures—assert the right or feel the obligation to make policy.

Decision making proceeds not by the flow of recommendations up and orders down, but by development of a shared sense of direction among those who must form the parade if there is going to be a parade. Collegial not command structures become the more natural basis for organization. Not command and control, but conferring and networking become the mandatory modes for getting things done. Planning cannot be done by a few leaders, or by even the brightest whiz kids immured in a systems analysis unit or a planning staff. Real-life planning is the dynamic improvisation by the many on a general sense of direction—announced by the few, but only after genuine consultation with those who will have to improvise on it. More participatory decision making implies a need for much information, widely spread, and much feedback, seriously attended—as in biological processes. Participation and public feedback become conditions precedent to decisions that stick.

That means more openness, less secrecy—not as an ideological preference but as a technological imperative. Secrecy goes out of fashion anyway, because secrets are so hard to keep. And policy widens out to become what Paul Appleby, that farseeing philosopher of public administration, called it a generation ago. "Policy," he said, "is the decisions that are made at your level and higher."[12] But note that his vertical language is already obsolescent.

Most of the history we learn in school is so narrowly focused on visible leaders that it may give us the wrong impression about leadership processes even in earlier times. We learn that Genghis Khan or Louis XIV or ibn-Saud or the emperor of Japan or George Washington said this and did that—as though he thought it up by himself, consulted with nobody, and wrote it without the help of a ghostwriter. But even in ancient, traditional societies I suspect that effective leadership consisted in being closely in touch with where the relevant publics were ready to be told to go.

Consensus is a prominent feature of many cultures now dismissed as primitive. The Polynesians in the Pacific Islands with their circular village councils and the American Indians around their campfires made (and in some degree still make) decisions by fluid procedures which may induce more genuine participation than a modern meeting run by parliamentary procedure. In the agora of Athens and the Roman "senate and public" (the SPQR), there seems to have been lively participation by those (wellborn male citizens) qualified to take part.

The difference in the current scene is the sheer *scale* of the relevant publics. In democratic Athens slaves, women, tradesmen, and other noncitizens did not presume to play in the decision games. The notion that all men, let alone whole peoples, had inalienable rights came in only with the Enlightenment, a scant three centuries ago—and has been made effective, still in a minority of the world's nations, only in the twentieth century. In Switzerland women *still* cannot vote.

Participatory fever is contagious. Public policy used to mean what the government does. Now it includes corporate policies, collective bargaining agreements, the cost of health care, the recruitment of university presidents, lobbying practices, equal employment opportunity, environmental protection, tax shelters, waste disposal, private contributions to political candidates, the sex habits of employees, or just about any other insider activities that outsiders think are important enough to engage their time and attention.

The biggest issues so far have to do with the quality of public responsibility that shows forth in the actions of corporations, universities, hospitals, and the thousands of other structures in which executives make the decisions that serve people, cost them, anger or please them.

The rising tide of participation is reflected in dramatic organizational changes. Big corporations now usually have a vice-president for keeping the corporation out of trouble with nosy outsiders, or even with their own stockholders and employees, who raise questions about what the company ought to produce, who it ought to employ, and how it ought to invest its money.

Should my company, or any American company, make and market nerve gas, even if the government does want to buy some? Should my company, or any American company, promote nuclear proliferation by selling to developing countries nuclear power plants that make plutonium, the fuel for nuclear weapons, as a by-product of generating electricity? Shouldn't my company have more women, and blacks, and American Indians in its employ—and especially in its board and top management? Should a company whose stock I own invest my money in South Africa? Should my company, or any American company, pass the social costs of its profit seeking—overcrowding, the paving of green space, radioactive risk, dirt, noise, toxic waste, acid rain, or whatever—to the general public? Should our community hospital perform abortions, splice genes, change people's sex, and invest in expensive equipment that can help only a few affluent patients? Should our state university do secret work for the Defense Department? Should the CIA recruit our students for who-knows-what clandestine wars in other people's countries?

Such questions cannot be brushed aside without raising their decibel level. There are ways to deal with all of them: shifts of policy, or consultative processes, diversionary moves, or public explanations—in descending order of probable effectiveness. But the visibly responsible leaders increasingly have to build into their organizations, not as a public relations frill but as an essential ingredient in bottom-line budgeting, staff members competent to help develop strategy on such issues as these. And the visible executive now has to be personally competent to defend the organization's public posture in public debate.

These public responsibility issues can make or break companies, products, and executive reputations. If you do not believe that, take a Nestle executive to lunch and ask him about marketing baby formula in the Third World.

INFLUENCE BASED ON SECRECY: DILEMMAS OF OPENNESS

The push for participation by all kinds of people and the inherent leakiness of the information resource combine to produce the modern executive's most puzzling dilemma. The dilemma must have been familiar to the first cave people who tried to bring other cave people together to get something done. But for us moderns, the scale of the perplexity is without precedent. The dilemma can be summarized in one question: How do you get everybody in on the act and still get some action?

The contemporary clamor to be in on the act is certainly impressive. In business, customers are feistier, more likely to complain; stockholders are more numerous and less passive; policyholders are more inclined to follow through on their insurance claims; union members and other citizens give advice on what is wrong with the steel and automobile industries; employees assert the right to judge whether their employers should make fragmentation bombs; maritime unions decide whether shipments should go to the Soviet Union; advocacy agencies excluded from the United Way organize their own competing drive for community funds; ethnic groups keep a watchful eye on investments in South Africa and business with the Arabs. More and more parents have a world population policy; teachers organize to tell school systems what ought to be taught; students want tailor-made courses of study. Environmental groups, carefully avoiding questions about whom they represent, are articulate (and effective) beyond the wildest dreams of Gifford Pinchot and Teddy Roosevelt. New kinds and colors of people are breaking through the oligopoly of influence long controlled by businessmen and male lawyers from early-arriving ethnic groups. Even those deadly predictable circuses, our national political conventions, become increasingly interesting as minorities and women fill more delegate slots and live TV coverage enhances the risk that a delegate will be seen making a deal, picking his nose, adjusting her shoulder strap, or falling asleep—in millions of living rooms at once.

Openness, then, is the buzzword of modernization. In its firmament the dieties are the public hearing, the news conference, the investigative reporter, "60 Minutes" and "20/20," Ted Koppel, Phil Donahue, and the *National Enquirer*. Its devils are also well known: smoke-filled rooms, secret invasions, hidden or edited tapes, and expense account luncheons at which The Establishment decides what to do next.

In consequence, compared with a generation ago most public officials—and a rapidly growing number of private executives conscious of their ultimate public responsibility—are much more inclined to ask themselves, before acting, how their actions would look on the front page of the *Washington Post* or the *Wall Street Journal* or on the evening telecast. Even former Vice-President Agnew has conceded that taking cash from contractors in his government office might be wrong if judged by what he called post-Watergate moral standards. No one doubts that raising the risk of public exposure will improve the private behavior of executive leaders as they ask themselves, "How would I feel about this action if everyone was able to see me take it?" The moral of Watergate is plain enough: If the validity of your action *depends* on its secrecy, better decide to do something else.

But the yen for wider knowledge and broader participation has gone well beyond this sensitivity training for visible leaders and has raised new questions about the cost-benefit calculation of more openness. A generation of experience suggests that it is high time we faced the *next* question: How much openness is enough?

Since this is not a mystery story, I will reveal at the outset the conclusion of the next few paragraphs. Experience teaches that the procedures of openness are well designed to stop bad things from happening and ill designed to get good things moving—unless the consensus for action has been built in private ahead of time.

A practical benefit of openness is simply that complex social systems work badly if they are too centralized. In managing their agriculture the Soviets have put this proposition on public exhibit for more than half a century. The opposite of centralization is of course *not* decentralization, which is simply an effort to preserve hierarchical workways when your organization gets too large for grandpa to know everything. The opposite of centralization is what Charles Lindblom calls mutual adjustment: in a generally understood environment of moral rules, norms, conventions, and mores, very large numbers of people are adjusting their behavior by watching each other and modifying their behavior just enough to accommodate the differing purposes of others, but not so much that the mutual adjusters lose sight of where they themselves want to go.[13]

What makes mutual adjustment work is the wide availability of relevant information, so each mutual adjuster can figure out what the others may do under varied conditions and give forth useful signals about his or her own behavior. The market principle does not guarantee smoothly working systems, of course; perfect competition among buyers and sellers with full information is to be found only in textbooks for sophomores. Yet very large systems, many of them global in scale, based on massive information outputs and feedback systems, have been developed in this century. In recent years systems unimaginable before the marriage of computers and telecommunications are accepted now as routine (currency and commodity markets, worldwide airline and hotel reservation systems, global public health controls and weather forecasting systems come readily to mind).

In other writings I have addressed the growing costs of openness. The very great benefits of openness and wide participation are flawed by oversimplification and confrontation, by apathy and nonparticipation, by muscle-binding legalisms, by too many meaningless public hearings, by an excess of voting and parliamentary process, by the nay-saying power of procedural objection, by the protection of mediocrity, by the inhibition of excellence in recruitment and the absence of candor in evaluation—and by one thing more. Mythology has it the other way around, but it seems clear now that wide consultation *early* in a policy process tends to discourage innovation and favor standpattism.

More openness in decision making is a radical litany, yet the multiplication of those consulted tends to water down radical reform. During the Vietnam war, I used to conduct seminars on this subject (among others) during the long hours spent with student leaders on the barricades. Why, I asked them, do you advocate openness with such passion when the reforms you want would be voted down if you put them to a big public meeting? They were regularly nonplussed by the question; evangelists, in David Riesman's phrase, often "mistake the righteousness of their cause for its marketability."[14]

An action proposal, especially if it is new and unfamiliar, will seem threatening, or at least postponable, to most of the experts who have not already been involved. It is no accident that so many memorable U.S. public policy initiatives (much of the New Deal and the Lend-Lease idea in the 1930s, the Marshall Plan and Harry Truman's Point Four in the 1940s, the Open Skies proposal in the 1950s, the Peace Corps and Food for Peace and the War on Poverty in the 1960s, the Nixon Doctrine and the Carter human rights initiative in the 1970s) began as the products of leadership hunch and thinking-out-loud rhetoric, with most of the professional staff work and the needed consultations at home and abroad following after. In each case the executive leaders were sensing a trend the American people would buy if a credible salesman came forward to peddle it. But if all the relevant experts had been asked for their opinions before launching them, some or all of these great ideas might well have shriveled in the womb. Too many people, in Washington and abroad, would have said, "Let's study it some more."

Bold initiatives for change can thus be killed by premature exposure to the rough winds of public debate. Yet let us again remember that this cuts both ways: timely openness is also well designed to stop foolish change. Earlier and wider consultation would almost certainly have killed the ill-fated Bay of Pigs operation, drastically modified the Vietnam escalation, and illuminated the grotesquerie at Watergate for the foolish scheme everyone, including President Nixon, later judged it to have been.

Whatever the costs and benefits of openness in particular cases, it is clear enough that in every kind of hierarchy the winds of openness and participation by new kinds of people are whistling through the cracks, blowing in the windows, and knocking down the doors. The result in each case cannot be so clearly judged in advance; it depends on how well those involved have analyzed and balanced the openness equation. Openness, like technology, is not properly an ideology. The answer to whether it helps or hurts basic human purposes is the same as the answer to most of the interesting questions in the study of society: it *depends*.

CLASS BASED ON OWNERSHIP: THE OBSOLESCENCE OF OWNERSHIP

The openness which the informatization of society brings in its train was bound to raise fundamental questions about the idea that knowledge belongs to a person or an organization—or a nation. The propensity of this sharing resource to leak is eroding the doctrine that knowledge can be owned, exchanged, and monopolized the way other resources can.

That you or I can own a fact or an idea, that a message of any kind belongs to a person or a corporation or a government, is (for reasons already cited from Colin Cherry's work) rather a peculiar notion to begin with. The person from whom you got the message did not lose it; any right you acquire by receiving it is at best shared with the sender, the carrier, and often a good many other nosy people who are privy to it. Even if you paid to get the message (if, for example, it was a piece of research you hired someone to do), or if someone paid to get it to you (a friend who sent you a cable, a company that sent you a commercial), it was the assembly or delivery service, not the information contained in the message, that was paid for. The researcher could not own the facts and ideas that she or he strung together for your use, and neither can you, even if you use them as your own.

The new tide of information technologies makes the ownership of intellectual property more detached from reality with every new invention. Dynamic high technology keeps developing better and faster *techniques of piracy*—xerography, videotape, the backyard dish for picking up signals from satellites. The knowledge explosion also produces new *kinds of works* (computer software), new *means of delivery* (microfiche, videocassettes, computerized videotext over a telephone line), and new *ways to assemble* great complexities of facts and ideas in more readily accessible form (computerized data bases, inventory controls, energy use data, on-line reservation systems for airlines and car rentals).

In this environment, laws written to protect books and phonograph records and broadcasts, the products of the past, are getting harder and harder to apply. Laws that address technologies not yet invented are hard to write. The nervous breakdown of copyright protection is now an open scandal. It may be retarded in degree by technological fixes. Satellite broadcasters can scramble their signals to prevent pirating. Elaborate codes have been devised by the creators of some computer programs, though teenage computer hackers have been showing how inherently porous they are.

When I first acquired a home computer, I found the ethical dilemma right up front: in the instruction manual. On its cover sheet I was threatened with litigious mayhem if I copied any of the software. On the very first page of the manual I was told that before I did anything else I should make at least two copies of the floppy diskette provided with the manual.

The makers of software keep up their pitiful efforts to maintain a proprietary interest in their products, but the happy-go-lucky free distribution of copies of copyrighted diskettes has already become one of the friendly gestures that makes the owners of personal computers feel like members of a new kind of guild. The leakiness of the information resource seems destined to overwhelm the backward-looking efforts to imprison it. The history of arms control, and the success of computer pirates, teach us that there is always a technological fix for a technological fix.

Is the doctrine that information is owned by its originator (or compiler) necessary to make sure that Americans remain intellectually creative? In most other countries creative work is overwhelmingly controlled by organizations and carried out by salaried people. In Japan even the most inventive employee is likely to have a lifetime job and receive salary raises in lockstep with his age cohort, his morale sustained not by personal ownership of his ideas but by togetherness in an organizational family.

Most U.S. patents are held by organizations (corporations, universities, government agencies), not by the inventors. Many copyrights, perhaps most, are held by publishers and promoters, not by the authors and songwriters the Founding Fathers may have had in mind when they sewed information-as-property into the U.S. Constitution.

An author or songwriter who helps a publisher make money should certainly participate in the proceeds. But direct agreements about profit sharing or joint venture arrangements (the movie industry is already full of relevant examples) seem a less fragile basis for such cooperation than the fraying fictions that the author owns the words in a book and that shared information is being exchanged.

In U.S. universities and research institutes, creative work is already rewarded mostly by promotion, tenure, and tolerant traditions about teaching loads and outside consulting. We generate a respectably innovative R&D effort in public-sector fields such as military technology, space exploration, weather forecasting, environmental protection, and the control of infectious diseases without the scientists and inventors having to own the ideas they contribute to the process. In the private sector, the leaders of industries on the high-tech frontier are already saying out loud that their protection from overseas copyists does not lie in trade secrets but in healthy R&D budgets.

The notion of information-as-property is built deep into our laws, our economy, and our political psyche—and into the expectations and tax returns and balance sheets of writers and artists and the companies, agencies, and academies that pay them to be creative. But we had better continue to develop our own ways, compatible with our own traditions, of rewarding intellectual labor without depending on laws and prohibitions that are disintegrating fast—as the Volstead Act did in our earlier effort to enforce an unenforceable Prohibition.

In international politics the notion that knowledge is owned by sovereign states is in maximum disarray. Every newly miniaturized recording or micrographic device and every new satellite launched for communication or photography or remote sensing makes it more difficult to sustain the doctrine that national governments can own, or even control, their information resources.

In 1979 the U.S. government sent two delegations to two world meetings about the control of information. At a UNESCO conference in Paris, the delegates righteously advocated the free flow of information, meaning information furnished by U.S. news agencies, U.S. television producers, and U.S. movie studios. A few weeks later, at the UN Conference on Science and Technology for Development in Vienna, an equally righteous group of Americans came out against the free flow of information, meaning information as a technology we were anxious to hoard. Both principles are authentically American: the right to choose and the right to own. In the international discourse, we will hardly be able to have it both ways. Yet there is no evidence that the two groups of delegates, and the government that instructed them both, perceived the irony or the contradiction.

The U.S. State Department, which instructed both delegations, seemed unusually disoriented by the new information environment when it ruled last year that Western European owners of IBM computers could not move them from, say, Birmingham to Manchester without first seeking U.S. permission. This assertion of extraterritoriality, over equipment produced by a multinational company with headquarters in the United States, was designed to prevent strategic equipment from flowing indirectly to Communist countries. Regardless of the merits of the case, the edict is simply unenforceable. In the global information society, the long arms of ownership and control are shrinking fast.

If information is inherently hard to bottle up, policies based on a long-term information monopoly are likely to have a short half-life. For the 1980s and beyond, the principle is clear: if the validity of your action depends on its continuing secrecy, do not depend on it.

In our generation-long arms race with the Soviet Union, successive U.S. administrations have managed to persuade themselves that each new U.S. weapons system (its made-in-America technology a continuing mystery to our adversaries) would enable us to stay ahead. In one of the most damaging of these actions, in the early 1970s, the

United States decided to stuff multiple independently targeted reentry vehicles (MIRVs) into single missiles. Despite elaborate secrecy on our part, the Soviets soon figured out how to do likewise. But since they (for other reasons) had built much bigger missiles boosted by more powerful rockets, they were able to stuff more MIRVs into their canisters than we could. Thus did we outsmart ourselves by taking an action that depended for its validity on technological secrecy, and create the famous window of vulnerability instead.

In the management of mutual deterrence the overclassification of information about what we could do, if we had to, may actually increase the danger of war by miscalculation. The core of the nuclear deterrent, that remarkably stable if unattractive substitute for peace, is the Soviet leaders' uncertainty about what the U.S. president would do in the event of Soviet moves against our allies or ourselves combined with their certainty that we have the means to retaliate no matter what. Keeping our intentions credibly uncertain is easy: we cannot know what we would do in a situation until we know what the situation is. But keeping from our adversaries full knowledge of our capabilities merely adds another element of madness to the nuclear arms race.

Our own government has for three decades engaged in halfhearted and demonstrably ineffective efforts to control strategic U.S. science and keep foreign nationals out of sensitive university research. In our mostly open society, it never worked very well. Americans have no corner on the market for brains; scientists talk quite freely across frontiers to each other; our European and Japanese allies never had much enthusiasm for controlling transborder information flows (because sales of equipment mean jobs for Europeans and Japanese); and Soviet technological espionage, like our own, has long been a thriving industry.

Keeping our R&D to ourselves is a policy that depends for its validity on secrecy. As informatization intensifies in the postindustrial world, strategic secrecy can be expected to work less and less well.

Similar government behavior used to work better for dictators and totalitarian bureaucracies in societies where keeping information from spreading is honored by doctrine and practiced ad absurdum. The last time I looked, Xerox machines still had to be licensed by the government in the Soviet Union; in Bulgaria, even typewriters are closely controlled. Ideas are harder to license: Russian youngsters readily learn about jeans and hard rock, and scientists on both sides of the porous Curtain seem to know how far along their peers are in unraveling (for example) the puzzlements of rocketry and space travel.

The good news is that information is leaky, that sharing is the natural mode of scientific discovery and technological innovation. The new information environment seems bound to undermine the knowledge monopolies which totalitarian governments convert into monopolies of power. In the horoscope of the USSR and the Soviet bloc, a future looms where nobody is in charge.

PRIVILEGE BASED ON EARLY ACCESS: EQUALITY OF ACCESS AND FAIRNESS

The informatization of society may destabilize more than the Soviet bloc. It may help undermine the systems that keep 2 billion people in relative poverty and more than a third of them in absolute poverty. In many ways the most exciting, and puzzling, question about the new knowledge environment is whether it will be good news or bad news for the global fairness revolution—and for that revolution's U.S. precinct, the upward mobility of women, minorities, and the poor.

The most arresting trait of the information resource is that it is inherently more accessible than other resources—and that, once accessed, it unlocks the other resources. What does that imply for access to the power and affluence that knowledge brings in its wake? Theoretically at least, compared to things-as-resource, information-as-resource should encourage:

- the spreading of benefits rather than the concentration of wealth (information can be more equitably shared than petroleum or gold or land or even water), and
- the maximization of choice rather than the suppression of diversity (the informed are harder to regiment than the uninformed).

In the industrial era, poverty was explained and justified by shortages of things; there just were not enough minerals, food, fibers, and manufactures to go around. Looked at this way, the shortages were merely aggravated by the tendency of the poor to have babies.

In the postindustrial era, the physical resources are joined at center stage by information, the resource that is harder for the rich and powerful to hoard. Each baby, poor or not, is born with a brain. The collective capacity of all the brains in each society to convert information into knowledge and wisdom is the measure of that society's potential.

But whether the informatization of the globe will actually mean a fairer shake for those who have been the victims of discrimination depends mostly on what they do. Most of the fairness achieved in world history has not been the consequence of charity, goodheartedness, and noblesse oblige on the part of those in power. Always in history, it seems, fairness has been granted, legislated, or seized when there was no alternative. And usually the reason there was no alternative was that those out of power were determined (or at least perceived by those in power to be determined) to cast off their shackles and take the law into their own hands.

Societies flexible enough to adapt to the pressure from groups out of power (as the United States has been doing, not without conflict and coercion, on school integration, voter rights, sex discrimination, and equal employment opportunity) manage to keep change comparatively peaceful.

Societies that try to maintain rigid hierarchies (and especially those which, like the Shah's Iran, at the same time encourage education for most of their people) get blown out of the water. The Shah of Iran was brought down by the marriage of convenience between two groups who harbored powerful resentments: mullahs who had been bypassed and downgraded by modernization were angered by the lack of respect for tradition, and Iranian students, at home and abroad, were angered by lack of fairness. Afterwards the tradition defenders and the fairness advocates went after each other, and the fairness people lost.

In other countries the mix is different, but part of the stew of resentments is always the complaint we learn from infancy to make: "It isn't *fair*."

There will be less excuse in the future than in the past for depriving whole populations of the benefits of development. There will also be less excuse for the disadvantaged to blame their condition on the barons and bosses when the accessible knowledge to even the score is already floating out there in the noösphere—the sphere of human consciousness and mental activity.

The noösphere is an *accessible* resource that has many of the characteristics of a commons. In considering the implications for fairness of information-as-a-resource, it is an intriguingly fresh thought, worth a moment of speculation.

In earlier times sharing arrangements for a common resource were customary, for example in tribal ownership and nomadic practices. Vestiges of the idea survive in the Boston Commons, the National Park system, and in the way many major waterways in Europe and North America are managed. For people in old England the commons, as Ivan Illich defines it, was "that part of the environment which lay beyond their own thresholds and outside of their possessions, to which, however, they had recognized claims of usage [not to produce commodities but to provide for the subsistence of their households]." The commons "was necessary for the community's survival, necessary for different groups in different ways, but . . . in a strictly economic sense, was not perceived as scarce."[15]

The older commons, such as those for sheep and cattle, have disappeared through enclosure. But the commons idea has now been revived in a big way, as the basis for worldwide cooperation in the environments that by common consent belong to no one or everyone (which seems to be about the same thing): the deep ocean and its seabed, Antarctica, outer space and celestial bodies, and the weather.

The Mediterranean Sea, the arena of bloody ancient feuds and lethal modern rivalries, has recently been formally recognized by all the coastal states (including the Arab states and Israel) as so precious a shared commons that reversing its degradation must be a matter for cooperation even among sworn enemies. The resulting international agreement, intermediated by the UN Environment Programme, is self-enforcing: violating its terms would be in a literal sense self-defeating.

For the management of an information commons, a sharing environment, these exotic precedents suddenly seem not so exotic. Illich, in a Tokyo speech called "Silence Is a Commons," argued that electronic devices (from the microphone to the computer) are a form of enclosure, reserving to the few the privilege of breaking the silence otherwise available to the many.[16] I do not know about silence; I have not had much experience with it. But on the computer as a form of enclosure, I demur. In its general impact the forced march of information technology, personal computers combined with global telecommunications, seems to me to be taking us away from the idea of enclosure. My hunch is that the fusion of computers and communications will further empower the many to participate in making policy in domains to which the few, with their moth-eaten monopolies of knowledge, will have to yield more and more access.

Neighborhood organizations are furnishing themselves with personal computers to deal more effectively with the banks, developers, and government agencies that will otherwise make the neighborhoods' decisions for them. American Indian tribes might set up a computer teleconference to concert their political clout on fast-moving legislation. A single individual with a personal computer can even now get access to so much useful and timely information that she or he can, with a week's homework and without leaving home, intervene as an unusually knowledgeable citizen in almost any public policy issue on the national agenda.

To chart these potentials is not to fulfill them. The trends in information technology would make it possible to organize as a commons (with free though not necessarily costless access thereto) most of the world's useful knowledge. That is not to say it will happen. It just helps remake the point that those who think "it isn't fair" will have plenty of opportunity to get access to almost any information that is being withheld from them to their disadvantage. But they will have to want to work at it, they will have to prepare their brains for the task. In the information society as in its predecessors, there is still no free lunch.

POLITICS BASED ON GEOGRAPHY: THE PASSING OF REMOTENESS

I have argued the mind-blowing implications of the informatization of society for four of the old hierarchies—based on control, secrecy, ownership, and structural unfairness. Let's look at what is happening to the fifth of the old hierarchies, those based on location.

The inherited idea is that the political importance of communities is based on their geography. Cities usually developed because they were seaports or on critical inland waterways, or (earlier) on important overland caravan routes and (later) on important railway lines. It made a difference whether you were in the city or in the country; if you lived in a rural area, you were remote. There was no choice about it, you were just remote.

The importance of countries was often based on the natural resources they had discovered, and developed, on their territory. The spices of the Orient, the rubber and tin of Southeast Asia, the coal and iron of Central Europe, the diamonds (and later, uranium) of South Africa, the fruit of Central America, the petroleum reserves of Indonesia and Mexico and Venezuela and North Africa and the Persian (or Arabian) Gulf and the North Sea, the soil that produced those "waving fields of grain" in North America—these crucial resources left an indelible mark on the national sovereignties which happened to find them in their backyards.

Then there was the sense of place in military strategy, summed up in the once-popular word geopolitics. This was the idea that a nation's power depended largely on its geography—how vulnerable its land mass, how defensible its frontiers, how rich its mineral deposits, how fertile its soil, how plentiful its water, how extensive its coastline.

But communications satellites and fast computers are gradually erasing distance, eroding the idea that some places are world centers because they are near other places or near obsolescent natural resources or near old-fashioned means of transportation, while other areas are bound to be peripheral because they are remote from these centers.

Octavio Paz, a poet, caught onto what was happening well before most of the systems analysts and political pundits. "We Mexicans," he wrote in the 1970s, "have always lived on the periphery of history. Now the center or nucleus of world society has disintegrated and everyone—including the European and the North American—is a peripheral being. We are living on the margin . . . because there is no longer any center. . . . World history has become everyone's task and our own labyrinth is the labyrinth of all mankind."

The passing of remoteness is one of the great unheralded macrotrends of our extraordinary time. Once you can plug in through television to UN votes or to a bombing in Beirut or a Wimbledon final; once you can sit in Auckland, or Singapore, or Bahrein and play the New York stock markets in real time; once you can participate in rule, power, and authority according to the relevance of your opinion rather than the mileage to the decision-making venue—then the power centers are wherever the brightest people are using the latest information in the most creative ways.

Distant farmsteads can, if they will, be connected to the central cortex of their commodity exchanges, their political authorities, their global markets. The fusion of rapid microprocessing and global tele-communications presents nearly all of us with a choice (and an obligation to choose) between relevance and remoteness. There will be costs and benefits to either choice—but the necessity to choose is new, and inescapable.

There is, of course, an alternative to geography as a principle of organization. The revised proposition was recently formulated by futurist Magda McHale: in the new knowledge environment, civilization will be built more around communities of *people*, and less around communities of *place*.[17]

That this trend is well advanced can be seen in a quick review of what is happening to the great geographic hierarchies which in this last couple of centuries have been dividing, and governing, the world.

The state is not withering away, as with their different motives Karl Marx and the advocates of world government would have desired. But power is leaking out of sovereign national governments in three directions at once. *The state is leaking at the top*, as more international functions require the pooling of sovereignty in alliances, in a World Weather Watch, in geophysical research, in eradicating contagious diseases, in satellite communication, in facing up to global environmental risks. *The state is leaking sideways*, as multinational corporations—private, pseudo-private, and public—conduct more and more of the world's commerce, and operate across political frontiers so much better than committees of sovereign states seem able to do. *The state is also leaking from the bottom*, as minorities, single-issue constituencies, special-purpose communities, and neighborhoods take control of their own destinies, legislating their own growth policies, their own population policies, their own environmental policies.

And what has nation come to mean? Increasingly it means not a hierarchy of power but ethnicity—the Frenchness of Quebec, the tribal loyalties of the Ibo in Biafra, the separatism of the Scots, the rhetorical brotherhood of the Arabs, the world's many diasporas, ranging from the Overseas Chinese to the Zionist and non-Zionist Jews outside Israel.

And organized religion? All of the great religious traditions have had to settle, so far in world history, for hegemony in one or another part of the world. But in a world of people-communities, not place-communities, the parish cannot be mostly geography-based. Now, even established religions are trying to break free from their national and regional parishes. The Roman Catholic pope's extensive travels and the terrorist outreach of Ayatollah Khomeini's Shiites form a grotesque correlation: both are breaking loose from historic geographic bounds to appeal to wider religious—and therefore political—constituencies.

The prospect of people rather than place as a basis for community has interesting implications for universities trying to serve a local clientele, for corporations that have bet heavily on regional organization, and for political systems that have bet heavily on geography-based constituencies. It implies that those institutions which exploit the electronic answers to remoteness may be catching a wave in the twilight of hierarchy.

SUMMARY AND CONCLUSIONS

In sum, the informatization of society will force dramatic changes in some long-standing hierarchic forms of social organization. The process of change, the slow accumulation of small changes in the way various social functions are performed, is far harder to discern than the ultimate result. Information technology pervades our lives and institutions in the same way that termites inhabit a house. As unseen termites consume the structural supports of a building, so may information technology challenge the rules, norms, and conventions that, in an earlier time, served to organize society by vesting economic and social power in centralized leadership, secrecy, ownership, resource control, and propitious geography. As with termite damage, we can be caught unawares at the collapse of those structures we thought sturdy, with the first visible sign of change serving also as proof positive that it is too late for stop-gap solutions. Therefore, if we are to avoid catastrophe, or a least avoid fighting the last war, we must both broaden and lengthen our vision of the future.

Information technologies will be assimilated without turmoil only if scholars recognize the need to rethink their disciplines in light of the erosion of societies based on material resources and industrial production. Citizens will have to get used to the responsibility that goes with the influence and power almost casually available to them through access to information. Generalist leaders will find themselves rethinking the nature of leadership. They more than others will have to widen their angle of vision to take in an informed and consulting public, and extend their focal length to take in the full implications of the twilight of hierarchy, not in the next 100 years but in their own life and work. We will need to change—we are already changing—the negligent procedures that left the Bomb unconstrained by hard thinking about causes and effects. The informatization of society holds great promise, but will need to be housed in a new intellectual home. The termites are at work on the old one, and we had best not wait until we lean on its wall and it caves in without notice.

NOTES

1. The statistics on the redistribution, between 1880 and 2000, of the U.S. work force were culled from the research of G. Molitor, Public Policy Forecasting, Inc., by Henry M. Boettinger, who headed AT&T's corporate planning before he joined E. F. Hutton as head of its Office of Information Strategy and Technology. "Information Industry Challenges to Management and Economics," *New Jersey Bell Journal*, Vol. 7, No. 1 (Spring 1984), pp. 12-21.

2. The OECD report, *Information Activities, Electronics and Telecommunications Technologies: Impact on Employment, Growth, and Trade* (Paris: OECD, 1981), was one of a series of reports OECD has conducted as it tries to trace the impacts of the Information Age on, and its implications for, member countries.

3. G. Edward Schuh's comments were made in a note to the author.

4. P. Drucker, *Managing in Turbulent Times* (New York: Harper & Row, 1980).

5. The first report to the Club of Rome, which sold more than 3 million copies in many languages, is available in book form as *The Limits to Growth*, by Donella H. Meadows et al. (New York: Universe Books, 1972). John McHale's book, *The Changing Information Environment* (Boulder, Colo.: Westview Press, 1976; first published in the United Kingdom in 1972) covers the changes in communications, resource use, education, business and management, and political process that derive from the impact of information upon society.

6. Dieter Altenpohl discussed the role of materials development and their social, economic, and ecological impacts (thus illuminating the role that information plays in materials substitution) in his book *Materials in World Perspective* (Heidelberg, Germany; Berlin: Springer-Verlag, 1980). His chart relating the weight of materials to their degree of sophistication, can be found on p. 201.

7. In his book manuscript entitled "A Second Industrial Revolution?," Professor Colin Cherry of the University of London explained the "sharing" nature of messages—

while making clear that *meaning* is not necessarily or even usually shared. The discussion here is based on that manuscript and on talks with Professor Cherry about that draft in Aspen, Colorado, the summer before he died.

8. Lewis Branscomb, "Information: The Ultimate Frontier," *Science*, January 12, 1979, pp. 143-147.

9. C. Cherry, "A Second Industrial Revolution?" (unpublished manuscript).

10. Comments about the obsolescence of both liberal and Marxist approaches are found in the influential examination of the effects of the new communication and computer technologies on society by Simon Nora and Alain Minc in *L'informatisation de la Société*. Originally produced as a commissioned report from the two civil servants to President Giscard d'Estaing, it quickly became a best seller in France and influenced the thinking of Giscard's successor, President François Mitterand. The book was eventually translated into English as *The Computerization of Society* (Cambridge: The MIT Press, 1981), with an introduction by Daniel Bell. The English version of the title misses the point that Nora and Minc were making in their book— which is that the *marriage* of computer and telecommunication technologies is the new dimension of society.

11. Chester Barnard's theory of executive process is spelled out in *The Functions of the Executive* (Cambridge: Harvard University Press, 1938).

12. Paul Appleby, *Policy and Administration* (University, Ala.: University of Alabama Press, 1949), p. 21.

13. Charles Lindblom's book, *The Intelligence of Democracy* (New York: Free Press, 1965), introduced the concept of "mutual adjustment" in the management of both public and private organizations and institutions. He updated his thoughts in a lecture at the University of Minnesota's Center for Strategic Management Research, "Incremental Strategy: Still Muddling Through," on May 13, 1983.

14. The comment by David Riesman comes from my extensive correspondence with him on the subject of openness in university governance.

15. Ivan Illich argued that computers are doing to communication what fences did to pastures, in "Silence Is a Commons," *The CoEvolution Quarterly*, Winter 1983, pp. 5-9.

16. See note 15.

17. Magda Cordell McHale, "The Feminist Model," Center for Integrative Studies, State University of New York at Buffalo, 1984.

Comments

ALEXANDER H. FLAX
President Emeritus
Institute for Defense Analyses

Harlan Cleveland's paper, as befits the subject of this symposium, is truly global in scope, not only geographically, but in its full sweep, in its contemplation of human affairs, and in treating civilization as an integrated whole. Its insights into the nature of changes in society being wrought by the rapid pace of progress in information technology are profound and thought-provoking.

It would be very difficult indeed to do justice to this paper in a brief discussion, but I think it would not be amiss to talk about a few things from the standpoint of an engineer, this, after all, being a symposium of the National Academy of Engineering.

It is heartening that Ambassador Cleveland sees the portents of the new information technology for society as predominantly benign. There are some doubts expressed, but mainly his is an upbeat appraisal, and this—for engineers who have lived through the traumas of the 1960s and early 1970s—is a welcome change. We all heard the dire predictions of unemployment that Norbert Weiner made and that many others picked up—you may recall *The Human Use of Human Beings*. We are once burned, twice shy.

We should not assume that society will naturally accept the benefits of new technologies without our doing anything about it. We must remain aware of those other voices saying that this new information technology may undermine civil rights in the United States through invasions of privacy and that the very openness of the information could result in pervasive data banks which could be used for repressive and harmful activities. We need a countervailing logic and this paper helps provide it, but I don't think it solves the problem for all time. In fact, there are some real problems that engineering can help solve in the handling of data to help prevent some of these abuses. We should always keep those in mind.

Let me also comment on the fact that Ambassador Cleveland has addressed, for the most part, the benign role of leaking information, or more pervasive information if you like. He has talked about information leaking primarily in the context of democratic societies. There are others who fear that in a totalitarian or police state some of the new information technology can be used to make repression more effective, much along the lines of Orwell's *1984* or Huxley's *Brave New World* or B. F. Skinner's *Walden 2*, and we must be aware of that dimension of the problem.

The other area that is of special interest to engineers is what the famous structural engineer Hardy Cross called the third member of the trilogy of engineering. The trilogy was, to paraphrase, science and technology, economics, and social dynamics. This was long before the protest movements, but nevertheless Cross counted the interaction of technology with society as very important. We cannot ignore the role of economics in striking an equilibrium. That is, any one of the elements of the trilogy cannot run wild because another one will check it, and that is as true for social wants, needs, and desires as it is for technology. We have seen that new information technology is subject to the inexorable laws of economics in the marketplace. You only have to recall the picturephone or the premature attempts to put computer-based instruction in the schools in the early 1960s. Neither took hold. I assure you that there will be more things that will not be accepted. That is how we discover what things work: we try them. Economics is going to put bounds and limits on how far we can go in any of these directions, although I think that we will go a long way and very much along the line that Ambassador Cleveland has described. Certainly this timely, wide-ranging, farsighted overview of where we are going with this new information technology deserves careful study by engineers.

Property Rights
in Information

ANNE WELLS BRANSCOMB

The subject of this paper—the laws of the Information Age and, more specifically, property rights in information—is a difficult one, but it is absolutely essential to pursue if we are to enter the Information Age knowledgeably and with an understanding of the consequences.

After an introductory section which presents definitions of property and historical background on information protection, the paper reviews recent developments with respect to property rights in information, analyzes several of the major areas of concern, and develops some general principles to guide us in the application of the law to the new technologies. For almost every right in this area there is an opposing claim or an adversarial relationship. Therefore, in each individual case it is a matter of balancing equities and sensibilities that often defy codification. As discussed in individual sections below, the rights include:

1. the right to know information about ourselves and the world we live in
2. the right to collect information—the investigative function
3. the right to acquire information—archived by others
4. the right to withhold information—about ourselves, personal, corporate, or national
5. the right to control the release of information
6. the right to receive compensation for information
7. the right to protect information—the security function
8. the right to destroy or expunge information
9. the right to correct or alter information

10. the right to publish or disseminate information—access to the marketplace of ideas

INTRODUCTION

Many scholars, including Colin Cherry,[1] Fritz Machlup,[2] and Harlan
Cleveland,[3] rightfully argue that information has characteristics different from those of natural resources and manufactured goods upon the
exchange of which our economic system rests. However, if we are to
transform our economy into one that relies primarily upon the economic
value of gathering, storing, processing, and distributing information,
we must develop principles from which we can derive economic value
for such activities. Therefore, it is not very helpful to the public debate
to insist that information must by its nature be shared or that it is
naturally leaky or uncontainable.

In civilized societies, especially in information societies that are
firmly rooted in an educated citizenry and intellectual prowess, we
will not tolerate the unnecessary spilling of proprietary information
any more than we will tolerate oil spills polluting our oceans. Neither
will we tolerate exclusivity with respect to information upon which
our livelihood as a nation depends. Inevitably we must turn to our
legal system to develop and to sustain those rights that we consider
inalienable and equitable and to delineate the boundaries between what
is considered public and what is to be protected by the law as private.

Definitions of Property

Property is a legal concept that dates back to the earliest history of
civilization and that is central to the efficient functioning of market
economies. The word deals with the boundary line between what is
yours and what is mine, or between what belongs to everybody and
what belongs to nobody.

Property, according to *Black's Law Dictionary*, is "that which is
peculiar or proper to any person; that which belongs exclusively to
one; in the strict legal sense, an aggregate of rights which are guaranteed
and protected by the government." The word is derived from the Latin
word *proprio* meaning to "own." The verb *appropriate* means "to
make a thing one's own; to make a thing the subject of property, to
exercise dominion over an object to the extent, and for the purpose,
of making it subserve one's own proper use or pleasure."

According to *Webster's*, the word *property* means, for tangible
objects, "something to which a person has legal title," and for intangible
rights, that "in which a person has a right protected by law." The

word *proper* itself is the root from which *property* is derived. *Webster's* defines *proper* as that "which is socially appropriate: according with established traditions and feelings of rightness and appropriateness," or that which is "sanctioned as according with equity, justice, ethics, or rationale," or that which is "marked by rightness, correctness, or rectitude . . . entirely in accordance with authority, observed facts, or other sanction."

Our Founding Fathers followed John Locke's definition of property as including "that property which men have in their persons as well as goods," and James Madison concluded: "In a word, as a man is said to have a right to his property, he may be equally said to have a property in his rights."[4]

In Spanish law the *propios* or *proprios* were certain plots of land reserved as the unalienable property of the town, for the purpose of erecting public buildings, markets, and so forth, or to be used in any other way, under the direction of the municipality, for the advancement of the revenues or the prosperity of the place.

In recent years international lawyers have become enmeshed in defining the ownership of international public spaces. Antarctica is the world's only landmass not territorially designated by proprietary ownership. However, the ocean seabed and outer space have been subject to much debate over what constitutes the "common heritage of mankind" or "the province of all mankind."[5]

Information is neither naturally proprietary nor naturally shared any more than the earth, or ocean seabed, or space is naturally the province of mankind or the property of individuals, nation-states, or other legal entities.

Although we have lagged behind the Japanese in our research and study of the social impacts of the Johoka Shakai (the Japanese term for information society),[6] we have led the world in litigation, legislation, and judicial interpretation of legal rights and obligations with respect to information. Public discourse for many years to come will concentrate on defining property rights in information that are marked by rightness and in accord with equity, justice, ethics, or reason, and will focus especially on defining those rights that are subject to an effective sanction.

History of Information Protection

The conflict between public and private information is as deeply rooted in our historical documents as is the protection of both private and public real estate. New England villages were built around a "common," or public area, in which villagers gathered on public occasions, much as the Spaniards promenaded around their central parks in the cool of evenings. In both cases the open spaces were essential to the exchange of information in communities that depended upon face-to-face voice communications. Counterparts of such places include the ancient Greek agora, or marketplace, and London's Hyde Park Corner.

We have protected our seacoasts for public access up to the high-water mark and have developed a great national park system for the protection of animals and for recreational activities. The concept of public ownership of airwaves arose at the time of Teapot Dome scandals when the public was outraged about the private exploitation of our great natural resources. There was no natural obligation to make the airwaves a public asset, as lawyers could map out private rights in spectrum resources just as easily as they have in land masses. It was considered the right and proper way of handling the Tower of Babel that existed on the airwaves and prohibited anyone from using that resource efficiently.

The Constitution provides for Congress to protect patents and copyrights, whereas the Bill of Rights in the First Amendment establishes an unregulated marketplace of ideas. Thus the Founding Fathers pursued the contradictory goals of protecting the work products of inventors and writers while at the same time preserving the public interest in the availability of information. We have tried to walk a tightrope between providing too much or too little protection for information. Cases abound in the law with regard to what constitutes "fair use" of information generated by another without his or her permission. The basic philosophy was that facts and ideas could not be protected, but only the composition (in the case of authors) and the embodiment of an idea in a product or process (in the case of patents).[7] Consequently, we have preserved the open access to laws of nature and to mathematical formulas that can be easily replicated. At the same time we have attempted to provide authors and inventors with legal protection for their expertise in order to compensate them for their efforts, to encourage greater productivity, and to increase the body of knowledge upon which human progress depends.

This concept of protection of intellectual property is not shared by all the nations of the world, nor is it derived from any natural law. Other cultures have their own concepts of propriety with respect to intellectual output. There is nothing universal about copyright, patents, or trade secrets. These are devices for reconciling the interests of society with those of the individual and of linking intellectual productivity with commercial gain. Their origins are in Western cultural history and are deeply rooted in the development of the printing press and industrial manufacturing societies. These are legal concepts designed to encourage a positive attitude toward innovation and social change. These concepts began to develop in the late Middle Ages; they have no logical counterpart in early civilizations where the products of intellectual expertise belonged to the community and not to their creators.[8]

All societies have developed an information policy, the most stringent of which is exemplified by the burning of books in China in 213 B.C. and the burying alive of some 460 scholars who were thought to be able to teach their contents from memory. Throughout history knowledge has been closely guarded, usually by a priesthood whose continuity depended on the maintenance of secrecy concerning rituals, herbs, hieroglyphs, or, in the case of American Indians, secrecy of the sand paintings used to heal the sick and bless newlyweds.

Thus we can consider computer hackers a *new* breed of priests, whose primary motive has been to disseminate information about new computer software as widely as possible rather than to keep their expertise closely guarded. It is a startlingly new concept, therefore— that of considering information as a natural resource which in its natural state will tend to permeate the society. Such dissemination has always depended upon skills that must be acquired, nurtured, and supported by social sanctions.

Knowledge can only be acquired by developing one's intellectual skills. Thus, the protection of intellectual activity developed alongside the rise of a class of individuals who had the leisure to produce artistic, literary, or useful inventions. There is substantial evidence that early Greeks recognized the rights of artists to be identified with their works and to have them performed or presented as created.[9] The first adjudication of a copyright issue seems to have taken place in the Middle Ages when a zealous clergyman visited his former teacher long enough to make a hurried copy of his Book of Psalms, the copying of which was not sanctioned by the king.[10] The elaborate rules that we are attempting to follow today awaited the advent of the printing press in 1436 and the rapid industrialization of Western Europe.[11]

The Constitution of the United States[12] endows the federal government with the power to regulate copyrights and patents as a device to "promote the progress of Science and useful Arts by securing for limited Times to Authors and Inventors the exclusive Right to their respective Writings and Discoveries." Thereby authors and inventors were encouraged to share their output with the nation. Surprisingly, there was little debate concerning this clause. The *Federalist* papers merely reflect that "the utility of this power will scarcely be questioned. The copyright of authors has been solemnly adjudged in Great Britain to be a right of common law. The right to useful inventions seems with equal reason to belong to the inventors. The public good fully coincides in both cases with the claims of individuals. . . ."[13]

In both cases it was assumed that it was in the public interest to encourage the widespread dissemination of knowledge and to nurture native intellect. Interestingly enough, the copyright law that was enacted gave protection only to American authors, denying any protection to imported works, which led to widespread copying of English publications.[14] This had the opposite effect from that of encouraging native talent allegedly intended by the legislation. As the piracy and copying of English information products were far cheaper than acquiring the original works of American authors, a flourishing publishing industry developed primarily for textbooks copied from English sources. Thus the Copyright Act of 1790, a high priority on the calendar of the First Congress, was a cornerstone in the philosophy of the new nation to encourage literacy and widespread dissemination of useful knowledge—an early development of technology transfer. It took another century to amend the act to recognize international copyright arrangements that gave mutual protection to American authors from unauthorized reprints by foreign publishers.[15]

Despite the long and complex historical experience with copyrights and patents developing protection for the production of information, it was only late in the last century that we began to build a body of law limiting the use of personal information.[16] In the last 90 years we have made great progress in accommodating ourselves to the intrusion into our personal and political lives of an independent press, both print and electronic media, and we are beginning to forge a new law with respect to computer communications as well.

THE RIGHT TO KNOW

The right to know can involve either simple or complex matters, e.g., the simple matter of the right of individuals to know of their origins and who their true parents are, or the more complicated matter of the right of the public to information that provides the basis for public decisions. However, no such right is absolute. For example, with the advent of amniocentesis, which permits identification of the sex of an unborn child during periods of pregnancy when an abortion can still be legally performed, it has become a controversial issue whether or not the couple should be told the sex of the child. Relatives and doctors withhold the information that a patient has a terminal disease where, in their judgment, it would be psychologically damaging to the patient.

Geraldine Ferraro and her husband confronted a difficult decision with respect to the release of information that Mr. Zaccaro considered would be damaging to his business interests. His tax returns were not legally required by the strictest interpretation of the campaign disclosure laws, but the public believed it had a right to know the facts in order to pass judgment on Ms. Ferraro's qualifications to be vice-president.[17]

"Anatomy of a Libel Suit," an excellent two-hour program produced for public television by the Columbia School of Journalism, dramatized the dilemma between the public's right to know and the private right to control personal information. It contrasted the role of the press in acting as surrogate for the public with the right of individuals to maintain secrecy about their personal affairs and the right of corporations to protect internal memoranda and documents that may prove embarrassing or damaging to their business. The rule of law that has evolved, as enunciated in *Times* v. *Sullivan*[18] two decades ago, is that the public's right to know takes precedence over the personal right to privacy. If one is a "public figure" (e.g., a public official—even a policeman on the beat) or a quasi or limited public figure (e.g., one who is well known in the press already with respect to the controversy in question), then the press can be held liable for misstating facts only where malice (e.g., intentional lying or flagrant disregard for the facts) can be shown.[19]

THE RIGHT TO COLLECT INFORMATION

The right to collect information is vested primarily in the government and is differentiated from the right of journalists to investigate the facts. Many businesses and institutions collect information—banks, insurance companies, credit institutions, hospitals, universities—but there is no legal right to do so. Most of the information is freely given or exchanged for services. In most cases there is a quid pro quo for the disclosure of the information based upon either an explicit or an implied contract. Having delivered sufficient consideration to legitimize a contract, the institutions consider the information proprietary and are loathe to release it even to the subjects of the information, be they patient, customer, or student. It has taken federal and state laws as well as court orders to require universities to make their records available to students and their parents,[20] and universities are still troubled about the consequences of having what they consider sensitive records open to inspection because it will inhibit professors and other staff from being candid about their observations. Also, a recent case holding that individuals have no inherent legal interest in records concerning themselves held by others is quite troubling.[21]

States as well as the federal government have been forging a new framework for privacy over the past several decades, and the effort has become more intense since the 1960s. The primary federal privacy legislation now in effect controls access of the public to government data banks in which information is collected and aggregated.[22] However, the right of the government to collect information is fundamental to the survival of the nation. Its economic health as well as its physical and political health is at stake. The census, which was started in 1890, is a fundamental function of the federal government. In order to collect demographic data and aggregate it for analysis, the government must have the power to compel citizens to respond. This is a very sensitive area, and many citizens do not agree that they should be compelled to release the necessary data. However, allocation of public funds is based upon percentages of identifiable groups, for example, children of the military living within a school district. Also, market information— important to a healthy private sector—is gleaned from the economic statistics that are collected by the Bureau of Economic Affairs in the Department of Commerce.

Stringent legislation exists to protect individuals from disclosure of information that is identifiable as specific to an individual or corporation. However, the access laws have made it extremely complicated to protect such anonymity for large enterprises, and many corporations feel that their trade secrets may be jeopardized even though one of the exceptions to the law is that such disclosure may occur.

A very disturbing decision was handed down recently by the New York State Court of Appeals. Under New York law domestic insurance companies are required to keep certain books and records, including minutes of their corporate board of directors' meetings. Such records have been voluntarily sent to the State Insurance Department under a promise of confidentiality. However, the court ruled that the agency must disclose the minutes of the directors' meetings to the *Washington Post* under New York's Freedom of Information Law, which defined "records" as "any information kept, held, filed, produced or reproduced by, with or for any agency . . ." and that an amendment to the act had eliminated any reference to an exemption for records "confidentially disclosed."[23]

THE RIGHT TO ACQUIRE INFORMATION

The right to acquire information is a public access issue that goes beyond the right of the press to investigate and/or the right of the government to collect information. It involves the right of citizens to acquire information already collected by others. James Madison observed that "a popular government without popular information, or the means of acquiring it, is but a prologue to a farce or a tragedy, or perhaps both." Based upon that verity, the United States has built its information policy upon a broad base of public education, public libraries, public subsidies for newspapers and other printed materials, and a government printing office that publishes the information gathered and processed by the federal government. Strictly speaking, public access may be difficult to conceive of as a property right. However, it is basically a legal entitlement to use public property, e.g., that information which is gathered and processed and published with federal funds and/or has become generally available to the public.

Patents and copyrights are the legal means by which authors and inventors are encouraged to share their creative endeavors with the public in exchange for the legally protected right to control the use of their intellectual products.[24] However, there has been considerable concern that many of these patents become buried in government archives and are never turned to productive use.

This concern is especially pertinent for patents owned by the government, where private enterprises have no incentive to develop the technology. Only 4 percent of the government's 30,000 patents have been developed and marketed.[25] In response to this concern, Congress amended the patent and trademark laws in 1980 to permit universities, small business firms, and nonprofit institutions to apply for patents on federally funded research and to retain exclusive licenses on these patents for as long as eight years.[26] Although this legislation has offered a cure for one problem—the loss of productivity from development of government patents—it has created another—the concern that the basic purposes of universities will be altered. In particular, there is concern that university efforts to sponsor research jointly with industry, in response to diminishing public funding of research, may change the public availability of information produced in universities.[27]

For the general purpose of making information available to the public, aside from the Government Printing Office (GPO), there are extensive federal libraries including the Library of Congress (LC), the National Library of Medicine (NLM), and the National Technical

Information Service (NTIS). These organizations collect and disseminate information developed with federal funds (as in the case of NTIS), registered with the Copyright Office (as in the case of the LC), or collected because it is pertinent to basic research being conducted by scientists and physicians both in federal employment and around the country (as in the case of NLM).

In recent years there has been much turmoil about the dissemination of information that has been processed by federally owned computers and is available in machine-readable form. The most recent policy enunciated by the present administration is that such information will be offered to private contractors for electronic dissemination. Users will pay for the direct cost of accessing the data from computer-based systems, but the federal government will exercise no influence over the fees that contractors or subcontractors may charge the public to access such on-line data.[28] Another pioneering project has been undertaken by the Trademark Office, which has entered into barter arrangements with private companies to exchange government data for services in designing the computer graphics software and entering the data to access trademark registrations on-line through an electronic data base. The ultimate goal is to produce the paperless office in which manual files will no longer be maintained.[29] This is a far cry from the previous practice of agricultural data being distributed freely and without charge by congressmen to their constituents and from that of county agents offering their consulting services to local farmers. Congressman Byron Rogers from Colorado's First Congressional District used to send baby books to all new parents with his compliments, and many useful publications are offered at a modest cost by the Government Printing Office.

Public libraries agonize over the high costs of distributing information on-line, as charges are not imposed on the the use of books unless they are overdue.[30] There is, of course, no natural rationale that requires public libraries to permit readers to read books in their collections without charge. Indeed, in former times libraries were very carefully guarded assets, access to which was strictly regulated. Neither is there a natural rationale which requires that access to computer-based information be subject to usage-sensitive charges. The main difference is that it is more convenient to measure and collect the marginal cost of computer access than it is for manual systems.

THE RIGHT TO WITHHOLD INFORMATION

The personal right to privacy is basically a right to withhold information from public dissemination or disclosure. It is quite limited in a society that believes in the free flow of information. In the United States the press is encouraged to serve as a watchdog and to seek out wrongdoing in public and private institutions.

As the press has become more and more diligent in its role and the means of surveillance and investigation have become more sophisticated, it is not surprising that public agitation with respect to the advent of new information technologies has sparked an interest in privacy laws. The most basic property right is the protection and integrity of one's own person—the right to withhold information about oneself that one considers to be one's own, the right to disclose information about oneself at such time and place and under such conditions as one chooses, and the complementary right to know of and to correct information about oneself that is inaccurate and damaging to one's pride or reputation.

Scholars of privacy insist that privacy is a natural instinct that has roots in the earliest of civilizations. However, when one visits the pueblos and cliff dwellings of the Southwest Indians it is difficult to believe that our early forebears on this continent enjoyed very much privacy. And those of us who grew up in small-town America also have serious doubts that privacy is so deeply rooted historically. Indeed, it seems a newly won privilege of urban civilization where the sheer numbers of citizens living together make it virtually impossible for everyone to keep track of everyone else. Therefore, it comes as something of a surprise that there is such commotion about information technologies that have the capability to reimpose the surveillance that was characteristic of small, tightly knit communities.

Nonetheless, it is a characteristic of our evolving civilization that we are developing an increasing respect for the individuality and privacy of every human being in addition to a recognition of proprietary rights in real estate and other material possessions. Therefore, it follows logically that we will also evolve a body of law to protect information about ourselves as well as information concerning our corporate enterprises and public institutions.

Some rights have become universally accepted and require no legal sanction. For example, many women in our society prefer not to tell their age, and that preference is usually respected. The privacy of other information, such as that about sexual behavior, is less universally accepted; e.g., we are only recently beginning to accept overt homosexuality among those in positions of power and influence such as the teaching profession and the judiciary. Consequently, there has been a strong incentive to protect such information from disclosure. Criminal activity, physical illness, psychiatric treatment, credit history, and sources of income are all areas in which we have developed concerns about disclosure if disclosure would be damaging to personal interests or if such information would be damaging to the public interest. A case in point was disclosure of the information concerning the psychiatric treatment of Senator Eagleton, which affected his viability as a candidate for the vice-presidency of the United States in 1972.

Trade secrets are protected under a very extensive body of law developed through state legislation and the common law. However, trade secrets law is full of contradictions and inconsistencies with the laws of copyrights and patents, particularly with respect to the protection of proprietary interests in computer software. The basic philosophy of both copyright and patents is to disclose information in order to expedite the public dissemination of information but to provide compensation to the authors or inventors, whereas trade secrets law rests upon nondisclosure to the public. Attorneys for computer software companies are in a great dilemma about whether to withhold, disclose, license, release, or rely upon royalties. To cover all bases is very difficult, since reliance upon one theory of law may foreclose use of another.[31]

Use of the social security number as a universal identifier has met with great opposition from proponents of privacy. Thus, many people refuse to disclose their social security number because they fear that the government will misuse the personal information referenced by social security number in many data banks. Nonetheless, social security numbers as identifiers have become widely used by both the government and private organizations. The majority of citizens seem to have no qualms about disclosing the number, and many feel that the advantages of cross-referencing data bases far outweigh the dangers.

Another area of the law that is very controversial and rapidly developing is the asserted right of journalists to withhold the identity of their sources. Here the right of the public to know is in conflict with the right of those accused of wrongdoing to know the source of the accusation. Investigative reporters claim that they will be unable to obtain information from those with knowledge, especially where their source's future may be jeopardized (e.g., by discharge from employment for "whistle-blowing" or by retribution at the hands of criminals for "squealing"). This area of the law is still very much in a state of flux.[32]

The Supreme Court, when reviewing a journalist's plea for a special privilege,[33] noted that Congress might elect to legislate a special privilege for those who inform the public, but it failed to restrict the category to journalists. The Court included lecturers, pollsters, novelists, academic researchers, and dramatists among those who might seek or merit the right to protect the confidentiality of their sources.

This right is greatly desired by most scientists, despite the seeming contradiction between the philosophical basis of an open exchange of information in the pursuit of knowledge and the closed system of individual sovereignty of the researcher over the management of the data, subject to a peer group review.[34] In 1971 Professor Samuel Popkin of Harvard University went to jail for a week for refusing to disclose the sources of his research related to issues in the Pentagon papers litigation.[35]

The only two areas in which scientific researchers can obtain protective orders to protect their sources are in cases of psychiatric illnesses or drug abuse.[36] Efforts to persuade Congress to enact the protection alleged to be required by scientific researchers in other areas have been unsuccessful.[37]

THE RIGHT TO CONTROL THE RELEASE OF INFORMATION

The corollary to the right to withhold information is the right to release information at a time or place of one's own choosing. This right is recognized for corporate products, and corporations carefully guard the release of information about new products. Similarly, news releases are dated, and publishers usually respect the embargo both as a matter of common practice and in order to preserve the relationship with the source of the information for future publication.

The right to control the release of information is particularly important to those engaged in scientific research. Traditionally (until recently), basic researchers neither sought nor received much financial reward for their discoveries, but they jealously guarded their rights of paternity. Dorothy Nelkin summarized their position as follows:

> Scientists resist external control as a threat to the quality and integrity of research and as an infringement on their right to control the production and dissemination of their work. From this perspective, the question of ownership is unambiguous: the concept of individual sovereignty guides scientific behavior.[38]

There is nothing more zealously guarded by scientists than the peer group recognition that comes from first publication of research results that may lead to an esteemed professorship or even a Nobel Prize. In his study of scientists, R. K. Merton concluded that a scientist's claim to intellectual property rights was limited to such recognition and esteem.[39] However, with the changing economic environment in which academic research is being conducted and with large profits to be claimed, especially in the field of genetic engineering, all this may be changing. The incentive to claim property rights in scientific research that extend beyond paternity is becoming more compelling.

Nelkin has astutely observed:

> This assumption [that scientific sovereignty is in the public interest] leads to a fundamental contradiction: the use of secrecy to maintain sovereignty within a community whose work is based on open communication of research findings.[40]

Interestingly enough, at the same time that the incentive to control research findings more efficiently has increased, the technological imperative is in the direction of making it easier and easier to share scientific research. Computer networking among scientists has become widespread, facilitating the cooperative efforts of scientists from many parts of the world in joint research projects. Such simultaneous and sequential exchanges in a dynamic electronic environment often make it difficult to determine paternity with any kind of legal validity,[41] and the bits and bytes flowing freely via satellite from laboratory to laboratory may make it difficult to trace what librarians call an "intellectual audit trail." Thus, the dilemma arises for scientists that the economic environment in which they now work impedes the optimum use of a technology that has the greatest promise for sharing their intellectual output with the rest of the world.

An additional problem has arisen with the public outcry to be able to see the information upon which the scientist's judgment is based rather than to rely upon the expert opinion of the research scientist as reviewed and validated by a peer group. This is the primary reason for controlling the release of scientific data. Research scientists check and recheck their data until there is reasonable assurance that it is reliable before sharing it with the court of peers who will judge its authenticity and validate it for public distribution. However, the data in a university computer may not be as easily protected from premature release as the research notebooks of a pre-computer period.

There are many other types of data bases where release of information provided by individuals is routinely made available to the public or even sold for profit by both private and public institutions. Have you noticed, for example, how quickly you receive solicitations from boating magazines and commercial houses after you have purchased a new boat and registered it with the state authorities or documented it with the Coast Guard? Direct mailers are notorious for the speed with which they sell your name to their colleagues after you purchase an item. All of this is expedited by the use of computers for recording and distributing addresses.

In most jurisdictions there is at present no legal right to require entities to which you disclose your name and address to protect that information. However, there seems to be a groundswell of increasing concern about such use.[42] A few states have acted to prohibit the use of motor vehicle registration lists for direct marketing without the consent of the registrant,[43] and a number of others have enacted legislation restricting the distribution of cable television subcription lists.[44] Most companies are able to acquire such information quite easily. Several companies specialize in the distribution of motor vehicle registration lists, which are considered to be the most valuable and up-to-date address lists. Some institutions, such as American Express, voluntarily inquire whether you wish to have your name distributed to other commercial enterprises, and the Direct Marketing Association promises to have your name removed from all mailing lists if you so request. However, a trial run of this system by the author merely precipitated a deluge of catalogs from many direct mail houses that had not previously known the address.

Perhaps in an open society whose economy rests upon commercial enterprise, some of us do not cherish the protection of our own names from disclosure to third parties without our permission. Indeed, the telephone company requires those who cherish privacy to pay extra for the privilege of an unlisted telephone number. However, it should be a prime tenet of the right to privacy that government agencies should not disclose personal information for commercial purposes without prior consent of the person.

Credit agencies freely exchange information about one's reliability. The legal status of such information within the care and within the files of another is not entirely clear. However, some recent cases are troubling. A bank was held legal owner of the information about a depositor, who had no right to enjoin release of that information.[45] Medical doctors are usually held to be the owner of their own files about patients, and attorneys are the custodian and rightful owner of their own files about their clients. The attorney-client privilege protects disclosure by the attorney of client information to third parties, and doctors operate under an ethical mandate not to disclose medical information about the contents of their files to outsiders. However, the ease with which electronic files are being invaded today suggests that the law is not yet in place that will protect such files in the Information Age and/or that neither public nor commercial enterprises are willing to pay the price of data protection by encryption.

The case of material obtained illegally is another instance in which the right to control release of information is important.   Where criminal liability is predicated upon such information, it has been a long-standing principle of U.S. law that information obtained by illegal search and seizure is not permitted to be used to convict.[46] However, recent Supreme Court cases suggests that this rule may be weakened.[47]

In the case of publication of illegally obtained information, the law is not so protective of the source, and the government was unable to obtain an injunction against the *New York Times* to prevent publication of the Pentagon papers.[48] However, the former—the case of material obtained illegally—rests upon a basic principle of fairness to defendants and the latter—that of publication of illegally obtained information— upon the overriding right of the public to know. Consequently, in each case there is a balancing of public against private interests in release of information. There is an implicit conflict between the interests of the First Amendment in open and uninhibited circulation of information and the rights of interested parties (whether individuals, corporations, or governments) to prevent the disclosure of information that they consider may be damaging to them.

The most interesting case to date in this area has been the Elizabeth Taylor action to enjoin ABC from televising a docudrama about her life in which she had enjoyed no participation nor had had any opportunity to review the contents. It was Miss Taylor's argument that the story of her life was hers to distribute.[49] As she was an actress of note whose family fortune rested upon her abilities as an actress and her resultant fame, it was a reasonable argument that she had not merely a personal right of privacy not to be invaded by publication of information about herself but that she had a commercial interest to be protected from exploitation by an unauthorized commercial enterprise.

Usually the legal claims in such cases are based upon libel, privacy, or false light, but Miss Taylor's claim was a more novel claim of misappropriation of information rightfully belonging to her. She based her claim upon statutory rights and the common law "right of publicity."[50]

Libel cases against producers of historical documentaries have been dismissed on the grounds that certain events are of such historical significance that the First Amendment confers "leeway" to permit the public to view the reconstructed portrayal.[51] The problem is more critical with living subjects, especially those like Miss Taylor whose livelihood is based upon the electronic presentation that is also the vehicle being used for the historical representation. Courts have held that the right to file suit for invasion of privacy would not be available to descendants, although "merchandising rights" have been held to be passed to subsequent generations.[52] Inevitably, in such cases there is a conflict of many interests: those of the authors to the integrity of their creative freedom, those of the public to view historical events in perspective and in the latest information technology, and those of the subject or the subject's heirs either to protect the integrity of information or to retain the right to receive compensation for distribution of information that was generated through the efforts of the subject. Rights of paternity and integrity have not traditionally been covered by U.S. copyright law, but are established by contract.[53]

THE RIGHT TO PROFIT FROM INFORMATION

There are many mechanisms for supporting the creation of intellectual products, such as the maintenance of monasteries, public universities, and venture-capital firms or the patronage of wealthy families, for example, in Italy during the Renaissance. Yet copyright and patent laws have been the mainstay of public policy in market economies since the Statute of Anne was enacted in 1709.[54] Basically, this first copyright law protected the right of authors to dispose of their works in return for financial remuneration rather than for public acclaim.[55]

As this system was designed for the print media and primarily for the protection of the financial interests of publishers rather than of authors, it has been the subject of much controversy in the electronic age. The law had great difficulty in accommodating itself to machines. Even as early as the days of the player piano, the courts had difficulty in finding that the instructions to the piano were a "copy" of the underlying work that could be perceived by the human mind rather than a machine.[56] The author John Hersey, a commissioner on the Commission on New Technological Uses of Copyrighted Works (CONTU), which advised the Congress on the applicability of copyright principles to computer software, wrote an anguished dissent to the CONTU recommendation that software be so considered:

> Works of authorship have always been intended to be circulated to human beings and to be used by them—to be read, heard, or seen, for either pleasurable or practical ends. Computer programs, in their mature phase, are addressed to machines. . . . The computer program communicates, if at all, only with a machine.[57]

Hersey's colleagues on CONTU did not have as much difficulty coping with computers as he did, and recommended to the Congress that computer software be absorbed within the rubric of copyright protection.[58] Indeed, even with doubts about the statutory protection, lawyers have shown great ingenuity in devising methods to protect the interests of their clients through licensing agreements, centralized royalty exchange programs, and contractual arrangements.

Computer Software—*Apple* v. *Franklin*

The arguments over legal protection for computer software have been extensive and have only temporarily been resolved with the enactment of a special amendment in 1980 to the Copyright Act of 1976. The amendment was specifically intended to authorize copyright law to cover computer programs,[59] although the Copyright Office had been accepting computer software for copyright registration since 1964.[60] The Copyright Office had, in fact, registered more than 2,000 such programs prior to the investigative hearings of CONTU, primarily offered by IBM and Burroughs, two of the largest developers of computer software.[61]

Much of the success of Apple in introducing the personal computer to the marketplace rested upon its successful efforts to encourage the free exchange of software programs by its users and the offer to facilitate the opportunity for circulation. However, when the Franklin Computer Company copied its operating program in order to facilitate the running of software designed for Apple computers on Franklin hardware, Apple went to court to enjoin the practice. There was no question about the source of the software, as the Franklin version even contained references to Applesoft. The basic legal argument was that object code, in which the Apple operating system was written, was not copyrightable. The lower court, in the spirit of John Hersey, was unable to cope with this "baffling" new technology and found no infringement. Where were the original "works of authorship" in this electronic gibberish? What was "fixed in a tangible medium of expression"?

The defendant also suggested that for the copyright laws to cover the situation there must be some representation of irreparable financial harm. Rather than harm to the plaintiff, the defendant argued that the irreparable harm would ensue upon issue of an injunction against Franklin, a fledgling personal computer manufacturer with only 1,000 machines sold, compared with Apple, the then leader in the field, with 440,000 computers sold and sales of over $335 million in 1981. Franklin argued that equities and public interest were on its side to foster competition in the industry and to keep a balance between competition and protection. However, the appellate court disagreed. Although finding that the underlying interest of the copyright law required a presumption of irreparable harm to the plaintiff, the court said the copyright law required no such actual showing. The court had no difficulty in finding that the $740,000 invested by Apple in developing its operating system software represented a substantial investment in jeopardy, which would satisfy the underlying public policy considerations. It also found no difficulty in obviating the necessity of discriminating between object and source code,[62] since one is used to more readily get access to the other. In effect the two satisfy the same functional purposes—to tell the computer what routines it is to perform. The Supreme Court agreed, or at least refused to enter the argument.[63]

However, the debate about the appropriate protection for software continues.[64] Some software firms, such as Freeware, offer their programs for a contribution on the theory that wide dissemination is desirable and, at least in their judgment, profitable. Other software houses, while attempting to protect their intellectual efforts through licensing agreements to control unauthorized copying, nonetheless admit that staying ahead of the competition in innovative programs is what protects their financial investment. It remains doubtful whether or not this will be true in the more stable economic environment when software programs have matured and found their natural niches in the marketplace. Indeed, some of the young geniuses who started their careers as computer hackers have admitted to being devastated by the piracy of their work product after hours and hours of labor expended in its creation. How public policy addresses this issue will remain of high priority in the Information Age, as economic interests become more and more tied directly to information products and services.

Videotape Piracy—*Sony Corp.* v. *Universal City Studios*

An even more difficult copyright issue, one that is no nearer resolution, is that of videotaping television programs off the air through the use of videocassette recorders (VCRs). As recently as the early 1970s, network programs did not carry a copyright notice, because the law presumed that broadcasting a program on television was not a "publication" but a private performance to be governed by the contract negotiations and agreements reached between the two parties—the network originator and the broadcasting station that aired the program. At the time there was no conceivable way a user could record and resell the performance. Consequently, there was no "copy" that could be the subject of copyright registration.

All that has changed with videotaping, which is flourishing. Some 4 to 5 million VCRs are now purchased annually[65] (13 percent of the

American public already owns VCRs[66]), and more than 14,000 videotape leasing outlets have cropped up all across America.[67] However, the receipt of royalties from the rentals of such recordings is prohibited by law under the doctrine of "first sale," which grants a copyright owner the right to transfer the ownership but not to control the secondary uses thereafter.[68] As a consequence, no royalties are collected on the millions of rentals of videotapes of popular motion pictures. The sale of such tapes has not been large, but the market is growing steadily. The release of *Raiders of the Lost Ark* just before Christmas in 1983 precipitated a record of 500,000 purchases in less than a month.[69] The rentals now reach close to a billion dollars annually and run about nine to one over purchases.[70]

Most of the litigation has involved the taping off the air of television distributed programming. The time to resolve the issue has taken almost a decade, as the initial suit was filed in 1976. Although the lower court did not see a violation of the Copyright Act in the at-home taping of television programs, the Ninth Circuit Court of Appeals agreed with the complainants that their rights had been infringed and that they deserved relief. The Supreme Court disagreed, admonishing the appellate court for enlarging the scope of "an article of commerce" that is not the subject of copyright protection and beyond the power authorized by Congress. In addition, the Court found that the widespread primary use of VCRs was for "time-shifting" or otherwise viewing a program that could not be seen at the time it was first televised. Thus, the respondent was unable to "show that the practice has impaired the value of their copyrights or has created any likelihood of future harm." Also, the Court determined that VCRs were used for other, quite legitimate purposes in addition to that of recording plaintiff's intellectual products.[71]

Taping of television programs for personal use appears to have become accepted as a fair use of copyrighted material. This is not in accord with the historical interpretation of fair use, since the programs are taped in their entirety. The use of the doctrine in the past has usually been restricted to copying portions of the work "for purposes such as criticism, comment, news reporting, teaching (including multiple copies for classroom use), scholarship, or research."[72] The rationale of the court must have been the unlikely efficacy of trying to put Pandora back into the box and the fact that no commercial use of the tapes was either alleged or documented. The case might go quite differently for the resale of programs taped off the air.

The holding was consistent with the photocopying case,[73] which had to reckon with the efficacy of trying to hold back a technology that had produced countless street-corner copying houses all over the country—a development that makes it possible for any citizen to become an information provider. Furthermore, the Williams & Wilkins case involved the photocopying by the National Library of Medicine and National Institutes of Health (NIH) of scientific articles (many of which were taken from the plaintiff's 37 professional journals) in single copies for distribution to the NIH staff of 12,000 plus satisfying thousands of requests from cooperating libraries around the country.

In each case the Court was concerned about dealing with a new technology in a judicial context rather than leaving the legislative function to Congress, but in both cases the justices were divided in their opinions.

The motion picture producers are now taking their case to Congress and seeking repeal of the first-sale doctrine. They are also pursuing their legal rights in the courts against all potential defendants in the videotaping industry but without any redress being sought against users whose only videotaping is for their personal use in the home.[74]

Insider Information—*SEC* v. *Winans*

Another concern of the public is that insiders not be permitted wrongful financial gain from the use of information prior to its disclosure to the public. The Securities and Exchange Commission (SEC) regulates the buying and trading of stock by the executives and owners of businesses who have available to them information that could affect the stock market price.[75] Historically, financial information has carried monetary value second only to that of military information. Indeed, the Rothschilds are supposed to have based their fortune upon advance information of the defeat of Napoleon at Waterloo brought to them by their fleet of carrier pigeons.

In an open society such as ours, there has been considerable ferment about manipulation of the stock market by those with sensitive jobs. The most recent case and the most innovative one is the indictment of R. Foster Winans, a former employee of the *Wall Street Journal* and columnist of "Heard on the Street," for using the advance information that he controlled for personal profit in stock market trading.[76] Indicted along with Winans were two others, another former *Journal* employee and a former stockbroker from Kidder, Peabody, & Company. Also expected to be indicted is an attorney who allegedly used the information and profited therefrom. Over a million dollars in profits were reputed to be gained by the four in the alleged conspiracy to defraud the readers of the *Journal* through a failure to disclose the financial interest of the writer in the outcome of his articles.

Although the indictment contains allegations of securities, mail, and wire fraud, it is this novel legal theory of a fiduciary obligation to his readers that has precipitated much speculation among legal scholars. Proponents of the First Amendment and defenders of the press such as Floyd Abrams deplore such disclosure of financial interest as a forced speech that is inherently abhorrent to the constitutional right of free speech. However, investors in the stock market who rely upon such reputable sources as the *Wall Street Journal* for their financial information must necessarily be concerned about the opportunity for manipulation of the data and the stock market. This case presents an interesting dichotomy for those who prefer an unregulated marketplace, since the integrity of the information must necessarily rely upon either the ethics of the profession or the rule of law. Certainly the dispatch with which Winans was released from his employment by the *Journal* underlines the concern of the paper for its reputation.

Clearly the SEC insider rules were not originally intended to anticipate such a situation. A basic sense of fairness to the stock-trading public suggests where the equities lie. However, the legal situation is not so clear. While the theory of a legally enforceable fiduciary duty has appeal to those concerned with fairness in the marketplace, it may not technically lie within the Supreme Court's restrictive definition of insider trading. Winans was not an officer of any of the companies in which he traded. He was in the position of gathering information about them and/or disclosing it at a time and place convenient to his trading posture. The most solid legal argument may be that he embezzled the information about the time of publication of his column from his employer and released it to his coconspirators, who became thereby participants in the alleged theft.

The most expansive exposition of the legal position of the courts was expressed in a 1968 case: "Anyone—corporate insider or not—who regularly receives material nonpublic information may not use that information to trade in securities without incurring an affirmative duty to disclose. . . . [Federal securities laws] created a system providing equal access to information necessary for reasoned and intelligent investment decisions."[77] Prior decisions have concluded that the fiduciary relationship necessary to bring the SEC rules into action required direct responsibilities to the stockholders.[78] Consequently, the government may lose the case on technical points. Nonetheless, the indictment itself is a major step forward in focusing public attention upon the problem of authenticating sources of information and protecting the integrity of information upon which the public must make decisions. Certainly, there will be amicus curiae briefs filed by representatives of the working press, and publishers will argue that government intrusion upon the self-regulation of the industry will do more harm than good. However, where financial interests affecting the buying public are concerned, I suspect that the scales will fall more heavily upon the side of regulation if the courts fail to protect the integrity of financial information. One caveat, however: When the state of Alabama enacted legislation requiring reporters covering state government to disclose their economic interests that might constitute a conflict, the reporters succeeded in obtaining an injunction against its use.[79] There was no presumption of wrongdoing as in the Winans case, since the dangers were prospective rather than retrospective. Certainly, this is one of the areas in which the divergent thrusts of the First Amendment draw a clear line of battle.

Other Cases

Another area for concern in the future is that of public officials commercializing their memoirs or using information from their official experiences for profit. However, public servants may argue that their "stock in trade" is their intellectual output during their incumbencies much as Elizabeth Taylor claimed a commercial interest in her professional persona and the right to exploit her name, image, and reputation. The difference, of course, is that public officials receive their financial support from public coffers, whereas performers are dependent upon commercial exploitation of their performances as the economic basis of their livelihood. The public policy question is whether news gatherers are derelict in their duties to the public when they pay their news sources for disclosure of information.

A more serious problem occurs in cases such as that of Watergate defendants Haldeman and Ehrlichman, who were paid for their appearances on television. Since Watergate, there is growing concern about the right of public servants to financial gain from their public service, or especially for their public disservice. However, there is very little case law on the subject. This is probably because there is no one to represent the public interest—no injured party to litigate the issue. Several decades ago litigation arose over the right of Admiral Rickover to claim a copyright in his speeches delivered during government service.[80] In another case, Richard Nixon was not permitted to claim executive privilege to block the subpoena of the special prosecutor for the famous incriminatory tape recordings.[81] More recently, the Ford memoirs became the subject of litigation when *The Nation* published excerpts or paraphrasings of some of Gerald Ford's observations about Watergate and its aftermath prior to the publication date scheduled and precipitating the cancellation of a contract with *Time* magazine to publish portions of the manuscript.[82] The court held that *The Nation*'s article—only 2,250 words, of which at most approximately 300 were copied—was a fair use of facts contained in the 200,000-word manuscript and were "the very essence of news and of history." After reviewing cases holding that neither news events, historical facts, nor biographical facts are copyrightable, the court summarized its view that public events cannot be monopolized by participants in them.[83]

> We have been asked to examine complex questions concerning the Copyright Act and the memoirs of a public official. Throughout our consideration, we have been guided by the conviction that the statute was not meant to obstruct the citizens' access to vital facts and historical observations about our nation's life. By far the greatest part of the article in The Nation was no more than the reporting of information concerning political decisions at the highest level of government. These facts were sown in and gathered from the shared ground of our country's history. They are the "property of all". . . .[84]

Inroads are also being made into the quite disreputable practice of some criminals committing their crimes for the purpose of gaining notoriety. Given the enthusiasm with which publishers pursue the right to publish such stories, it is not surprising that we have begun to think twice about the consequences of promoting financial gain from anti-social acts as we become more dependent upon information products for our economic health. A prudent and wise rule, it could be argued, would not permit a criminal to profit from his or her own wrongdoing. Proceeds should more equitably be allocated to the victims or their families or be used to reimburse the government for the cost of incarceration. The former has been addressed in the enactment of what are called the Son of Sam laws. These were precipitated by the chain of murders committed by David Berkowitz in New York several years ago.[85]

THE RIGHT TO PROTECT INFORMATION

There are many areas in which we have recognized a right to protect sensitive information that has already been gathered, processed, and archived. For example, the dockets of juvenile offenders are not available for public inspection in many jurisdictions. Medical records are not open to the public. Government documents are sometimes stamped TOP SECRET, SENSITIVE, or EYES ONLY. Yet all of these protected areas are suspect under the opposing theory of the public's right to know and the fear that activities that take place in secret and that are not recorded in publicly available documents will cover up actions that are not in the public interest or that reflect negligence or incompetence.

Sunshine Laws and Freedom of Information Laws

Sunshine laws and freedom of information laws[86] have wreaked havoc with the traditional practices of organizations to conduct their business in private and to disclose actions to the public after debate is over. Universities as well as industry smart under pressure to open up their inner circles to greater representation from the public. Seats on college and university governing boards now include student members as well as faculty and staff, and corporations seek board members with public constituencies.

All of this is a very healthy and inevitable trend in an open and vital information society. It may encumber decision making and render it more costly, but democratic governance demands broad participation in making public policy. Such participation necessarily requires widespread dissemination of information upon which public decisions are predicated.

Strategic Information

However, sunshine laws and freedom of information laws also present new obstacles to security agencies. Gathering of strategic intelligence by tradition is assumed to operate in seclusion beyond the reach of public surveillance. Many people are concerned that the new technologies make it even more difficult to protect information of a sensitive nature because of the ease with which invaders can break into the system. The movie *War Games* is a good example of the fear that has been engendered concerning the potential for harm by unauthorized entry into computerized defense systems. Yet advances in computer cryptography have swung to the side of those who encrypt, if they use sound methods. We are assured that breaking into the data bases of the Department of Defense and the Central Intelligence Agency and National Security Agency is far more difficult than is represented by Hollywood. Still, it is a worrisome problem both from a technical and from a legal standpoint. Recently officials were astonished to discover that wiretapping a voice line was prohibited by statute, yet tapping into a digital transmission was not.[87] The law lags far behind the technology.

Other countries are more concerned than ours is about the vulnerability of information that is gathered, processed, and/or maintained in data bases outside their own boundaries. This is not entirely a specious argument, as the assets of Iran were impounded during the Iranian hostage crisis. Also, the billions of dollars whirling daily through the SWIFT computerized network from bank to bank internationally are a good example of information of high value that cannot by its nature be controlled physically within the confines of a single nation. A country that tried to do so would become isolated from the international economy and would have to fall back upon its own resources.

This has not inhibited countries such as Sweden from elevating the question of national vulnerability to a high level of public anxiety.[88] In the case of Sweden the concern was prompted initially by discovery that the Malmö fire department processed its data in a computer time-sharing service in the United States. To be invaded by aliens and to discover that your strategic data were stored in an enemy computer would certainly put an end to a war prematurely. As a result of such fear, nation-states purchase and maintain computers for their own essential information rather than use the more expeditious and less costly route of time-sharing with other countries. However, this is the price of maintaining national sovereignty over critical information in the absence of any international protocols for such protection outside their own geographical boundaries, within which they are presumed to have absolute jurisdiction.

Export Control of Technical Data

The current argument over new legislation to replace the old Export Administration Act,[89] which has now expired, highlights the problems that ensue when technical data are treated as a controlled commodity. Despite the long-standing existence of the restrictive clauses contained in the earlier versions of the act, it has only recently come to the attention of scientists that much of their existing open practice concerning the exchange of unpublished scientific data, in person, by mail, by conference, and by computer, is permitted only by the notion of a General License by the Department of Commerce.

The Bucy Report of the Defense Science Board[90] recommended that the export control system should reduce the number of controlled items on the Controlled Commodities List and concentrate on the transfer of applied technologies, the mastery of which would permit hostile nations to replicate U.S. military manufacturing capability. Thus, they recommended restricting only "arrays of design and manufacturing knowledge," and "keystone" manufacturing equipment. However, although the Congress incorporated the board's definitions into the 1979 act, it had left no guidelines concerning how such determinations were to be made. The extensive proposals of the Department of Defense, such as the highly controversial and secret Militarily Critical Technologies List, have not yet been approved by Congress.

Because the proposed new legislation permits U.S. law to reach out and control companies in other friendly countries to which technology has been transferred, the European Economic Community (EEC) has filed formal complaints through diplomatic channels deploring this intrusion upon national sovereignty and noting that the economic sanctions of the U.S. government over the Soviet pipeline had led to considerable "political and commercial damage."[91]

Transborder Television—Protection of Cultural Identity

Another example of protective policies has been initiated in Canada, which is concerned more about cultural identity and its own economic independence than about strategic security, since the main threat it suffers from the United States is to its cultural and economic integrity. In certain respects Canada has been ahead of the United States in entering the Information Age. Marconi's first transatlantic radio transmission was from Nova Scotia rather than from Truro, Italy. Morse's telegraphy found uses in Canada before it came into use in the United States, and Canada orbited domestic communications satellites two years before the United States was able to "get off the ground." Furthermore, Canada has been years ahead of the United States in its anticipation of the social impact of the new technologies and in integrating public policy towards the use of new technologies into national priorities.

One of the primary concerns of the Canadian Department of Communications has been the spillover of U.S. television stations across the Canadian border and the widespread development of cable television to pick up such stations.[92] As a consequence of these nearby sources of prime-time entertainment, Canada is one of the most heavily cabled countries in the world. The first reaction of the Canadian authorities was to propose not licensing cable systems that undertook to retransmit foreign broadcasts.[93] However, the public response to this proposal was so overwhelming that it was never put into action. The next effort was intended to stop advertising dollars (estimated at $20 million to $25 million annually) from going to U.S. television stations rather than their going to the Canadian Broadcasting System. Canadian authorities in 1971 determined that Canadian cable systems should delete the U.S. advertising from their retransmissions of U.S.-originated television programs and substitute Canadian advertising instead.[94]

This brought an uproar from border broadcasters who took their fight into Canadian courts and began to agitate in the United States for retaliation. The Canadian Radio-Television Commission (CRTC) suspended further implementation of the commercial deletion policy, but initiated as an alternative a rule that Canadian companies could not deduct the cost of advertising on non-Canadian communications facilities.[95] This added fuel to the long-standing controversy between U.S. and Canadian authorities that has yet to be resolved. Variously, the U.S. Congress has tried to implement retaliatory measures that would prompt the Canadian authorities to relent. An effort has been made to enact reciprocal legislation, although there are no genuine economic sanctions that would be comparable. Efforts have been made to restrict the deductibility of U.S. attendance at conventions in Canada and to restrict the marketing of the Telidon videotex system in the United States. A complaint filed with the Special Trade Representative[96] resulted in a finding that the Canadian practice was unreasonable, burdensome, and restrictive to U.S. commerce. Although both Presidents Carter and Reagan have called for mirror-image legislation, no practical solution has yet been reached.

This Canadian border spillover problem is a good example of a nation-state's determination to protect information that it considers proprietary and to exclude information that it finds objectionable. However, it is also evident that the two countries have widely divergent philosophies underpinning their information policies—the United States is dedicated to a free marketplace for information as protected by the First Amendment and best achieved through an unregulated economy, whereas the Canadians look upon information policy as a means to an end—the promotion and protection of their own cultural identity and economic viability.

THE RIGHT TO DESTROY OR EXPUNGE INFORMATION

Just as the Chinese more than 2,000 years ago resorted to book burning to remove the collective memory of prior social systems, governments today use less stringent but similar methods of changing the course of history. It is usual to call attention to the Soviet practice of rewriting history with a Marxist interpretation, but we are not as aware of efforts within our own society to do the same thing. Frances Fitzgerald reports her findings that textbooks used in American schools have changed their views of history on the average every five years since the 1930s.[97] Little Black Sambo is no longer black, and the 1930s version of *King Kong* with its fearful monster finally overcome has turned into a 1970s version with the multinational oil company exploiting the beast and destroying the island culture by depriving the natives of their religious symbol. Each reflects the social concerns of its own generation, and such reinterpretation of fiction is accepted as literary license. Aside from a very real ethical concern about accuracy, objectivity, and reliability of information, every society will accommodate its information policy to its own political circumstances.

There are sound reasons for destroying some records when they have served their purposes. Shredding machines would not sell so well were it not so. In a society where the printed word is pervasive and copies abound, it is not so easy to destroy all copies. However, within a centralized data base where it may be more efficient to keep the source material on file and merely make copies at the convenience of the user, we may need to worry about what is to be expunged and when in order to avoid information overload or overspill and/or to get rid of redundancy. With computerized archives being so roomy and so cheap, it may become uneconomic to clean out the files. In addition, with so many coauthors and coinventors, it may be difficult to determine who has the right to expunge or delete.

THE RIGHT TO CORRECT OR ALTER INFORMATION

Public Records

As important as the right to acquire public information is the right of individual access to personal information. This was the consensus of data protection leaders from nine European countries attending a 1984 conference in Bellagio. The mere existence of such right, they concluded, is likely to have a salutary effect upon keepers of records.[98] Although many European countries have centralized governmental authorities who license data bases and act as ombudsmen to enforce the rights of citizens, the United States has no counterpart. Such rights as are available are left to the courts to enforce. These include the right to compel the correction of inaccurate or untrue data in government files.[99]

Broadcasting

The broadcasting counterpart of the right of the individual to access personal information in public records is the personal attack doctrine, which permits the subject of an attack to obtain a copy of the offensive audio or video broadcast and to reply.[100]

Newspapers

There is no right in the print media, comparable to the personal attack doctrine, that permits an individual to reply to material already published. In order not to chill the freedom of speech that the First Amendment seeks to protect, the appropriate recourse to an inaccurate or untrue statement is a libel action after the fact for money damages. However, in California, where many libel actions are filed by prominent entertainers, the libel action may not be pursued if the publication voluntarily published a retraction of the alleged inaccurate information.[101] Also under the rubric of privacy protection (a competing First Amendment right), an injunction may be obtained to prevent publication of material that will place the plaintiff in a "false light."[102]

Efforts to establish news councils to serve as watchdogs on such abuses and to urge offenders to publish replies have not been very successful, as some of the major news companies refuse to participate. The effort of the Florida legislature to enact a right of reply for newspapers was struck down by the Supreme Court as intrusive upon the editorial judgment of the news media.[103]

Consequently, the new technologies of both broadcast and computer have come under more stringent rules than the old media. However, it must be remembered that the abuses to be corrected at the time of the First Amendment were strong government intrusions into news content, whereas today the greater fear is the danger of private corporate control over the means of distributing information.

Many of the major newspapers and magazines regularly publish corrections of items from previous editions. They should continue to do so in their own interest and that of the public. As they enter the age of electronic publishing (and many, such as Gannett, Time, Inc., and Dow Jones, are already delivering their content via satellite or engaging in the delivery of videotex), they may become subject to the laws governing the electronic media. However, they may find that the public is more sympathetic with the later evolution of the law as a more equitable and workable solution to balancing property rights in information in the sophisticated technological environment. The First Amendment shield, with which they so vociferously assert their right to be protected from public accountability, may not serve them well unless they have themselves conformed to the tenets of its underlying philosophy.

THE RIGHT TO DISSEMINATE INFORMATION

None of the other rights would be very meaningful if there were no right to send forth the information into the marketplace of ideas. However, such dissemination carries an economic cost when any technology other than the human voice is employed. Every community has some place that operates as a public forum. In small-town America it was around the steps of the county courthouse. Portuguese sailors who settled New Bedford met on street corners, and some New England villages meet at the town dump. Most societies have large squares in which great numbers of citizens can assemble for public festivals and pronouncements of policy—the Mall in Washington, D.C., Red Square next to the Kremlin in Moscow, St. Peter's Square in the Vatican, Red Square next to the Forbidden City in Beijing, and Marcos Square in Manila. Lafayette Square in Washington, D.C., may have once served that purpose, but it has long since been landscaped with gardens and walks that discourage protestors from marching near the White House.

However, the technologies of print, broadcast, and computer have made this method of disseminating information obsolete except as a backdrop for television versions that can be disseminated to many millions more than could ever be accommodated in one place.

In the United States we accommodated our policies very rapidly to the print media by establishing a system of post roads over which the mails and the newspapers could travel. We have provided preferential rates to books, newspapers, and magazines, and even to merchants' catalogs—and a special franking privilege to congressmen for expediting messages to their constituents. A nationwide penny postcard made it possible for the constituents to reply with low cost, and even to send their messages to each other.

The applicability of this principle of a public forum carries over into a concept of common carriage for voice messages on telephone and telegraph. The law has had great difficulty, however, in coping with the broadcasting industry because content decisions and carriage media are both under the control of a regulated broadcaster. Nonetheless, there is a limited public forum responsibility in the fairness doctrine which requires that broadcasters devote time to opposing viewpoints when issues of public importance are aired.[104] Numerous efforts have failed to establish a right to communicate over the broadcast media for the purpose of responding to commercials on the basis that they were really "infomercials"[105] or to obtain a right of access to the electronic media for the initiation of public issues.[106]

The courts have been very protective of broadcasters and only recently reaffirmed the long-held view that television is not a public forum even if it is controlled by a public agency.[107] Viewers have no right to compel any particular kind of expression over the public airwaves. The Constitution requires content neutrality only with respect to an activity that has traditionally functioned as an open marketplace of ideas. However, the Supreme Court did not view public television as assuming this role.

In renouncing the Federal Communications Commission's (FCC's) aborted attempt to establish public-access channels on cable television, the Supreme Court concluded that the FCC was attempting, without legislative authorization, to create a public right that did not exist, and it commented that the right to speak does not include a right to an audience in a nonpublic forum such as newspapers, magazines, or on the Senate floor.[108]

However, the power of the press, as a surrogate of the people, is a powerful tool for reform. As the movie *Gandhi* so vividly portrayed, the presence of the *New York Times* reporter and that paper's publication of Gandhi's fight for India's independence gave international credibility to the separation movement, increasing its influence internally.

For the protection of human rights the exposure of violations to public view may be the only way to impose sanctions upon the wrongdoers. The student demonstrators and ghetto youth of the late 1960s understood this better than did public authorities. As Arthur Clarke so eloquently expounded concerning his brainchild, the geostationary orbital communications satellite:

> The very existence of the myriads of new information channels, operating in real-time, will be a powerful influence for civilized behavior. If you are arranging a massacre, it will be useless to shoot the cameraman who has so conveniently appeared on the scene. His pictures will already be safe in the studio five thousand kilometres away; and his final image may hang you.[109]

Certainly, both the access of the news gatherer or researcher to the subject matter and the availability of a means of distributing the information are essential to the health of the public enterprise in a democratic society. How to assure this access in an economy that puts high priority both on using private businesses to provide the technological means of access and on noninterference with freedom of speech is a quandary of utmost importance to the legal profession and the public.

SOME OBSERVATIONS ON PRACTICAL PRECEDENTS AND PHILOSOPHICAL PRINCIPLES

The newer information technologies have created some options and opportunities that transform the environment for interactive communications. They have also precipitated much soul-searching concerning basic principles that should apply to their use. However, these technologies do not come to us in a legal vacuum. As a nation we have many legal theorems and practical precedents gleaned from a rich heritage of concern about basic rights in information in other media to guide us.

We can, with some assurance, make several observations:

1. We are reluctant to reward wrongdoers for their wrongdoing.
2. We prefer to encourage and reward innovation even through novel legal theories based upon notions of common sense, commercial fairness, and human justice.
3. The courts will not do a useless thing like trying to curtail the videotaping of broadcast material in private homes or to prohibit the photocopying of published copyright materials for personal use.
4. The courts will not try to expand a legal theory that is ridiculous in its application, e.g., attempting to differentiate between source and object code as a means for determining copyrightability of computer software.
5. The courts are reluctant to apply novel legal theories that have not received the sanction of the administrative agency which has jurisdiction over the subject matter, e.g., enunciate a judicial First Amendment right of access to the broadcast media.
6. The law must conform to public acceptance of what is right and equitable.
7. The concept of property rights, whether applied to material goods or to intangible information, is neither simple nor absolute. It reflects the values of a society that equally cherishes individual freedom of action and the sharing of information for the common good.

It is also possible to derive some basic principles that underpin our philosophy of property rights in information.

1. We must have a public medium through which information may be exchanged freely between information providers and information users. This may be a common-carrier channel (for telephone, telegraph, or electronic mail), a public-access channel for cable television, an op-ed page in newspapers, or a public bulletin board such as the Democracy Wall in the People's Republic of China. The village green or public

auditorium simply will not suffice in the age of electronic information highways.

2. Liability for content should rest upon the originators and producers of information, not on the carriers, unless they are one and the same.

3. The sources of information and the nature of their economic interest must be appropriately identified and authenticated in order to establish credibility and to permit those harmed by a message to seek redress, unless there is some overriding public interest in protecting the identity of the source who may be personally harmed or put in jeopardy by the disclosure.

4. Proprietary rights in information generated for commercial purposes should be recognized, legally protected, and compensation provided through justiciable means.

5. The piracy, embezzlement, misappropriation, or misuse of information should be punished.

6. Freedom of speech should not be impaired except in circumstances in which there is a clear and present danger that the health and safety of the nation is in question or that the health and safety of an individual is in serious jeopardy. This should be true whether or not the person in question is a public figure or private person.

CONCLUSIONS

The basic principles outlined above need not be circumvented in an electronic age any more than they have been in the past.[110] They are derived from a long legal history of concern about the integrity and independence of the individual in a free society. What is needed is a commitment to preserve the principles we hold dear in a digital environment no less than in a voice circuit or on a printed page. The late Ithiel de Sola Pool, who devoted his last intellectual effort to the preservation of First Amendment principles in the Information Age, wrote:

> The mystery is how the clear intent of the Constitution, so well and strictly enforced in the domain of print, has been so neglected in the electronic revolution. The answer lies partly in changes in the prevailing concerns and historical circumstances from the time of the founding fathers to the world of today; but it lies at least as much in the failure of Congress and the courts to understand the character of the new technologies. Judges and legislators have tried to fit technological innovations under conventional legal concepts. The errors of understanding by these scientific laymen, though honest, have been mammoth. They have sought to guide toward good purposes technologies they did not understand.[111]

These good and well-intentioned public servants need the sound judgment and sage guidance of scientists and engineers if the law is to make reasonable sense in the Information Age. It is important that there be mutual understanding among those involved in both legal and technological innovations as these developments clash at the crossroads of change.

NOTES

1. C. Cherry, *A Second Industrial Revolution?* (unpublished manuscript).
2. F. Machlup, *The Production and Distribution of Knowledge in the United States*, Princeton University Press, Princeton, N.J., 1962.
3. H. Cleveland, "Information as a Resource," *The Futurist*, pp. 34-39, December 1982.
4. "Legal Lore," *New York State Bar Journal*, pp. 49-50, May 1984.
5. *Compare* art. I of the *Treaty on Principles Governing the Activities of States in the Exploration and Use of Outer Space, Including the Moon and Other Celestial Bodies*, 18 U.S.T. 2410, T.I.A.S. 6347, 610 U.N.T.X. 205, signed at Washington, London, and Moscow on January 27, 1967, and entered into force on October 10, 1967, with the Preamble of the *Information Composite Negotiating Text of the United National Third Conference on the Law of the Sea*, A/Conf. 62/WP, 10 Rev. 2, done at Caracas on April 1, 1980, not yet in force and as yet unsigned by the United States.
6. Y. Ito and K. Ogawa, "Recent Trends in Johoka Shakai and Johoka Policy Studies," *Keio Communication Rev.* 5:15ff, March 1984.
7. A. R. Miller and M. H. Davis, *Intellectual Property: Patents, Trademarks and Copyright*, pp. 18-19, West Publishing Co., St. Paul, 1983.
8. E. W. Ploman and L. C. Hamilton, *Copyright: Intellectual Property in the Information Age*, pp. 4-9, Routledge & Kegan Paul, London, 1980.
9. These rights can be traced down to the so-called moral rights of "paternity" and "integrity" contained in art. 6 of the Paris Convention of 1971 (the latest version of the Berne Convention). U.S. copyright does not protect these rights, and the opposition of Hollywood producers to them may explain the failure of the United States to ratify the Berne Convention. *See generally* R. Brown, *Kaplan and Brown's Copyright*, p. 656, Foundation Press, Mineola, N.Y., 1978.
10. Ploman and Hamilton, *op. cit.*, *supra* note 8, at 8.
11. *Ibid.*, at 9.
12. Art. I, sec. 8, cl. 8.
13. B. Ringer, "Two Hundred Years of American Copyright Law," in *Twenty Years of English & American Patent, Trademark & Copyright Law*, p. 117, American Bar Association, Chicago, Ill., 1977.
14. *Ibid.*
15. Copyright Act of March 3, 1891, 26 Stat. 1106. *See generally* Ringer, *op. cit.*, *supra* note 13, at 127.
16. S. Warren and L. Brandeis, "The Right to Privacy," *Harvard L. Rev.* 4:193, 1890.
17. Dow Jones News Service documents 120827-0219, August 24, 1984; 120822-0348, August 21, 1984; 120821-0312, August 20, 1984.
18. 376 U.S. 254 (1964).
19. B. W. Sanford, "Twenty Years of Actual Malice," *2 Communications Lawyer* 1, Summer 1984. The definition of "public figure" has been greatly expanded beyond that of public officials by the recent reversal of the $2.05-million judgment awarded

by a jury in the libel suit of William Tavouleareas, president of Mobil Corporation. Tavouleareas unsuccessfully argued that if he had become a "public figure" it was because of notoriety brought about by the libel and was not due to any inherent public function of his position. In announcing his intent to appeal his case to the Supreme Court, he said, "I am appealing because I believe the law must not accord one institution in our society the unrestrained power to so damage our leaders that it jeopardizes our society's ability to function." *Communications Lawyer 1(3)*:10, Summer 1983.

20. *See, e.g.*, The Family Educational Rights and Privacy Act of 1974, P.L. 98-380 sec. 513, 20 U.S.C.A. sec. 1231g, and the Texas Open Records Act, sec. 7 of art. 6252-17a, V.T.C.S., and Office of the Attorney General of Texas, Open Records Decision nop. 229, October 26, 1979.

21. U.S. v. Miller, 425 U.S. 435 (1976).

22. The Privacy Act of 1974, P.L. 93-579, December 31, 1974, 88 Stat. 1896, Title 5 U.S.C. sec. 552a, as amended P.L. 94-394, September 3, 1976, 90 Stat. 1198, P.O. 95-38, June 1, 1977, 91 Stat. 179.

23. In the matter of the Washington Post Company v. State Insurance Department et al. No. 73, State of New York, Court of Appeals, March 29, 1984.

24. In the case of copyrights for a period of the lifetime of the author plus 50 years, 17 U.S.C. sec. 302(a); in the case of patents for a period of 17 years, 35 U.S.C. sec. 154.

25. "New Patent Bill Gathers Congressional Support," *Bioscience 29*:281, May 1979.

26. The Patent and Trademark Amendment Act, P.L. 96-517, December 12, 1980.

27. *See generally*, D. Nelkin, "Proprietary Secrecy Versus Open Communication in Science," *Science as Intellectual Property: Who Controls Scientific Research?*, pp. 9-30, Macmillan, London and New York, 1984.

28. Department of Agriculture RFP 84-00-R0-6, March 15, 1984. According to remarks reported in *Commerce Business Daily*, February 28, 1984, the Office of Management and Budget considers this RFP a prototype for distribution of electronic data by the federal government. Examples include *Market News Reports* from the Agricultural Marketing Service and *Situations Reports* from the Economic Research Service.

29. The authorization for this project contained in P.L. 96-517, sec. 9, requiring development of a comprehensive plan for transferring the files to an electronic data base, and P.L. 96-247, Title 35 U.S.C. sec. 6, authorizing "cooperative exchange ventures," are being challenged by the "cottage industry" of trademark searchers who use the hard-copy files without paying fees for access.

30. This was not always the case. The first public library in the United States was started by Benjamin Franklin in Philadelphia in 1731 as a subscription library. There were 50 original members, who paid 40 shillings to join and 10 shillings per annum. Subscribers paid double for books not returned. It was not until 1800 that the Library of Congress was started, and the New York Public Library opened in 1837. M. C. Tyler, "The Historic Evolution of the Free Public Library in America and Its True Function in the Community," in B. Taylor and R. J. Munro, eds., *American Law Publishing 1860-1900*, Glanville, Dobbs Ferry, N.Y., 1984.

31. D. M. Davidson, "Protecting Computer Software: A Comprehensive Analysis," *Jurimetrics Journal 23(4)*:339 at 400f, Summer 1983.

32. The leading Supreme Court case in this area is Branzburg v. Hayes, 408 U.S. 665 (1972), which recognized the necessity to compel disclosure to a grand jury when

criminal behavior is involved. *See also* 99 A.L.R. 3d 37 and P. L. Glenchur, *Hastings L. J. 33*:623-652. Courts will not countenance tortious behavior in the gathering of news. *See* Galella v. Onassis, 487 Fed. 986 (2d Cir. 1973) where defendant was enjoined from approaching the plaintiff closer than 25 feet.

33. Branzburg v. Hayes, *supra* note 32.
34. For an excellent discussion of this issue, *see* D. Nelkin, chap. 4, "Rights of Access Versus Obligations of Confidentiality," in *Science as Intellectual Property: Who Controls Scientific Research?* Macmillan, New York and London, 1984.
35. United States v. Doe (In re Popkin), 460 F. 2d 328 (1st Cir. 1972), *cert. denied, sub nom.* Popkin v. United States, 411 U.S. 909, 1973.
36. Public Health Service Act, as amended 1974, 42 U.S.C. sec. 242(a) and the Comprehensive Drug Abuse, Prevention, and Control Act of 1970, P.L. 95-633, 21 U.S.C. sec. 242a(b) and sec. 872(d).
37. *E.g.*, Privacy of Research Records Bill, S. 867 (April 4, 1979) and H.R. 3409 (April 3, 1979), 96th Cong. 1st Sess.
38. Nelkin, *op. cit.*, *supra* note 27.
39. R. K. Merton, ed., *The Sociology of Science*, p. 273, University of Chicago Press, Chicago, Ill., 1973.
40. Nelkin, *op. cit.*, *supra* note 27.
41. Pool & Solomon, "Intellectual Property and Transborder Data Flows," *Stan. J. Int'l. L. 16*:113, 1980.
42. The Direct Marketing Association has been monitoring some 80 pieces of proposed legislation during the last several years.
43. *E.g.*, New Jersey, Pennsylvania, Nevada, and Virginia.
44. *E.g.*, Illinois, California, Connecticut, and Wisconsin.
45. *Op. cit.*, *supra* note 21.
46. Miranda v. Arizona, 384 U.S. 436 (1966); *see generally* 30 A.L.R. Fed. 824. For cases concerning illegal beepers, bugging, and wiretapping, *see* 57 A.L.R. Fed. 646, 59 A.L.R. Fed. 959, and 97 L. Ed. 237. *See also* M. Goldey, "Aspects of International Voice Communications to and from the United States," *Jurimetrics J. 24(1)*:8-12, Fall 1983, regarding electronic surveillance of international mts calls outside the United States and the admissibility of such evidence in courts.
47. Minnesota v. Marshall, ___U.S. ___, 79 L. Ed. 2d 409 (1984); New York v. Quarles, ___U.S. ___, 81 L. Ed. 2d 550 (1984); Berkemer v. McCarty, 468 U.S. ___, 82 L. Ed. 2d 317 (1984); Massachusetts v. Sheppard, 468 U.S. ___, 82 L. Ed. 2d 737 (1984).
48. U.S. v. New York Times, 403 U.S. 713 (1971).
49. Elizabeth Taylor v. American Broadcasting Companies, Inc., 82 Civ. 6977 (S.D.N.Y. 1982).
50. New York Civil Rights Law, sec. 50-51; the Lanham Act for damage to protectable service and trademarks, and unfair competition under New York General Business Law, sec. 368(d).
51. *See* Street v. NBC, 645 F. 2d 1227 (1981), *settled and cert. dismissed*, 70 L. Ed. 2d 636 (1981).
52. *Compare* Hicks v. Casablanca Records, 464 F. Supp. 426 (S.D.N.Y. 1978) and Maritote v. Desilu Productions, Inc., 345 F. 2d. 418 (7th Cir. 1965) with Lugosi v. Universal Pictures Co., Inc., 172 U.S.P.Q. 541 (Cal. Super. 1972) and Price v. Hal Roach Studios, Inc., 400 F. Supp. 836 (S.D.N.Y.).

53. *See, e.g.*, Gilliam v. American Broadcasting Companies, Inc., 538 F. 2d 14 (2d Cir. 1976).
54. 8 Anne, c. 19, republished in R. S. Brown, *Copyright*, p. 851, Foundation Press, Mineola, N.Y., 1978.
55. Ploman and Hamilton, *op. cit., supra* note 8, at 30.
56. White-Smith Music Publishing Co. v. Apollo Co., 209 U.S. 1 (1908). This inadequacy has been cured by the 1976 act, which defines a "copy" as anything tangible from which the author's work can be replicated. 17 U.S.C. sec. 102(a).
57. *Final Report of the National Commission on New Technological Uses of Copyrighted Works*, 27-31, Library of Congress, Washington, D.C., 1979.
58. *Ibid.*, at 12; subsequently enacted 17 U.S.C. sec. 101, 117, as amended P.L. 96-517, sec. 10, 94 Stat. 3028.
59. Computer Software Copyright Act of 1980 ; Act of December 12, 1980; L. No. 96-517, sec. 10; 94 Stat. 3015, 3028; 17 U.S.C. sec. 101, 117.
60. Office of the Register of Copyrights, Announcement SML-47 (May 1964); Copyright Office Circular 31D (January 1965).
61. CONTU Final Report, *op. cit., supra* note 57, at 85.
62. "Computer programs are the ordered set of instructions which can operate a computer. . . . Source code can be written in languages which are English-like, such as BASIC or FORTRAN. . . . Source code instructions are either directly used by a computer or are first translated into the computer's machine language as 'object' code. Object code is usually printed as ones and zeros, but can also be printed as octal numbers (0-7) or hexadecimal numbers (0-15), with A-F representing decimal (10-15). Object code can be directly translated into 'assembly' language, in which machine instructions are represented by mnemonics. . . . Object code, the direct symbolic representation of the machine language, is intelligible to trained engineers" (e.g., like the piano player that was only readable by experts). D. M. Davidson, "Protecting Computer Software: A Comprehensive Analysis," *Jurimetrics Journal 23*:339, 341 (Summer 1983).
63. Apple Computer, Inc. v. Franklin Computer Corp., 545 F. Supp. 812 714 F. 2d 1240 (3d Cir. 1983), *cert. den.*, 104 Sup. Ct. 690 (1984).
64. *See generally* the excellent article by Duncan Davidson, *op. cit., supra* note 31, which discusses the various methods of protecting software and the concerns of lawyers about the viability of each.
65. Dow Jones News Service, Doc. no. 120118-0360, January 7, 1984.
66. Dow Jones News Service, Doc. no. 120706-0544, July 6, 1984.
67. Dow Jones News Service, Doc. no. 110512-1159, May 12, 1983.
68. 17 U.S.C. sec. 109(a).
69. Dow Jones News Service, Doc. no. 120119-0661, January 19, 1984. Until that release *Flashdance* and *Jane Fonda's Workout* had been the top-selling videotapes, with 200,000 copies each. *See also* "Hollywood Thriving on Video-Cassette Boom," New York Times, Monday, May 7, 1984, pp. A1, C17.
70. Dow Jones News Service, Doc. no. 110512-1159, December 5, 1983.
71. Sony Corp. v. Universal City Studios, ___U.S. ___(1984), 104 Sup. Ct. 774, at 778.
72. 17 U.S.C. sec. 107.
73. Williams & Wilkins Co. v. United States, 420 U.S. 376 (1975).
74. Dow Jones News Service, Doc. no. 110512-1159, May 5, 1983.
75. Rule 10(b)(5).

76. *New York Times*, August 29, 1984, pp. A1, D4; *New York Times*, Sunday, May 27, 1984, pp. F1, F21.
77. S.E.C. v. Texas Gulf Sulphur Co., 401 F. 2d 833 (2d Cir. 1968).
78. *See* Chiarella v. United States, 455 U.S. 222 (1980), involving an employee of a financial printing house who decoded documents about mergers and acquisitions; S.E.C. v. Dirks, *cert. granted*, 459 U.S. 1014 (1982) involving a stockbroker; U.S. v. Newman, 664 F. 2d 12 (2d Cir. 1981) involving employees of Morgan Stanley who traded shares in a takeover target represented by their firm; S.E.C. v. Thayer (pending) involving friends of LTV Corp. CEO who may have benefitted from passing confidential information to his personal friends; and S.E.C. v. Brant (the civil case against Winans and his friends). For a discussion of all of the above cases, *see* L. Wayne, "Inside Trading by Outsiders," *New York Times*, May 27, 1984, pp. F-1, F-21.
79. Lewis v. Baxter, 368 F. Supp. 768 (D.C. Ala. 1973).
80. Public Affairs Press, Inc. v. Rickover, 369 U.S. 111 (1962). The case raises interesting questions about the copyrightability of the speeches, i.e., whether they were government documents that cannot be copyrighted under Title 17 U.S.C. sec. 105, whether they were already in the public domain because of their oral delivery and circulated copies, or whether they were the private utterances of a public official in a nonofficial capacity. On remand it was decided that the speeches were delivered in the admiral's "private capacity" and that the fact that they were typed, duplicated, and cleared by the Navy was irrelevant, 268 F. Supp. 444 (1967).
81. Nixon v. Sirica, 487 F. 2d 700 (D.C. App. 1973).
82. Harper & Row, Publishers, Inc. and the Reader's Digest Association, Inc., v. Nation Enterprises and the Nation Associates, Inc., 723 F. 2d 195 (1983).
83. Time, Inc. v. Bernard Geis Associates, 293 F. Supp. 130 (S.D.N.Y. 1968); International News Service v. Associated Press, 248 U.S. 215 (1918); Hoehling v. Universal City Studios, Inc., 618 F. 2d 972 (2d Cir.) *cert. den.*, 449 U.S. 841 (1980); Rosemont Enterprises, Inc. v. Random House, Inc., 366 F. 2d 303 (2d Cir. 1966) *cert. den.*, 385 U.S. 1009 (1967).
84. 723 F. 2d 195.
85. "Fifteen states have passed 'Son of Sam' laws freezing proceeds from moneymaking ventures such as book sales of those locked up for capital crimes until claims by victims of their survivors are satisfied." *Christian Science Monitor*, April 5, 1983, p. 1.
86. The Government in the Sunshine Act, P.L. 94-409, September 13, 1976, 90 Stat. 1241, Title 19 U.S.C. sec. 420, Title 5 U.S.C. sec. 551f., and the Freedom of Information Act, P.L. 89-487, July 4, 1966, 80 Stat. 250, Title 5 U.S.C. 552, as amended, P.L. 90-23 sec. 1, June 5, 1967, 81 Stat. 54, P.L. 93-502 sec. 1-3, November 21, 1974, 88 Stat. 15-61.
87. 18 U.S.C. sec. 2511 arguably might not include such transmissions.
88. Commission on New Information Technology, *New Views: Computers and New Media—Anxiety and Hopes* (1979). *See also* J. Freese, "The Vulnerability of Computerized Society," *Transnational Data Rep. 4(5)* at 21 (1981).
89. 50 U.S.C. sec. 2402 *et seq.*, as amended.
90. Defense Science Board Task Force on Export of U.S. Technology, *An Analysis of Export Control of U.S. Technology—A DOD Perspective*, Office of the Secretary of Defense, Washington, D.C., 1976.
91. Dow Jones News Service, Doc. no. 110325-0319, March 24, 1984.

92. For an excellent discussion of these transborder issues, *see Cultures in Collision: A Canadian-U.S. Conference on Communications Policy*, Praeger, 1984, esp. chaps. 3 and 6.

93. CRTC Public Announcement, *The Improvement and Development of Canadian Broadcasting and the Extension of U.S. Coverage in Canada by CATV*, Ottawa, December 3, 1969, p. 1.

94. CRTC Public Announcement, *The Integration of Cable Television in the Canadian Broadcasting System*, Ottawa, February 26, 1971. *See also* CRTC Policy Statement, *Cable Television: Canadian Broadcasting: A Single System*, Ottawa, July 16, 1971.

95. Canadian Bill C-58, An Act to Amend the Income Tax Act, September 1976.

96. Under sec. 301 of the Trade Act of 1974. The complaint was filed on August 29, 1978, and hearings were held in November 1978.

97. F. Fitzgerald, *America Revised*, Atlantic Little Brown, Boston, Mass., 1979.

98. *Transnational Data Report*, vol. vii, no. 4, p. 195, June-July 1984.

99. Privacy Act of 1974, 5 U.S.C.S. sec. 552a(g) and (1) (A); *see also* R.R. v. Department of Army, 482 F. Sup. 770 (D.C. 1980).

100. 28 U.S.C. sec. 315, (a); 47 C.F.R. sec. 73.123, 73.300, 73.598, 73.679.

101. Sec. 48a of the California Civil Code n2 provides in part: "1. In any action for damages for the publication of a libel in a newspaper, or of a slander by a radio broadcast, plaintiff shall recover no more than special damages unless a correction be demanded and be not published or broadcast, as herinafter provided. . . ." The California court hearing the Carol Burnett libel suit against the *Enquirer* cleared the way for the record $1.6-million verdict by ruling that the publication was a magazine not a newspaper.

102. Seattle Times v. Rhinehart, 82-1721, is on the Supreme Court docket to determine whether the Washington State Supreme Court was correct in upholding an injunction for defamation and invasion of privacy by the leader of a religious group who sought to suppress publication of information obtained during preparation for trial, *National Law Journal*, October 17, 1983, p. 5. Injunctions are more often sought to protect so-called merchandising rights or the right to publicity. *See, e.g.*, Haelan Laboratories, Inc. v. Topps Chewing Gum, Inc., 202 F. 2d 866 (2d Cir. 1953) *cert. den.* 346 U.S. 816. "This right of publicity would usually yield them no money unless it could be made the subject of an exclusive grant which barred any other advertiser from using their pictures."

103. Miami Herald v. Tornillo, 418 U.S. 241 (1974).

104. 47 U.S.C. sec. 315 (1976), *aff'd*, Red Lion Broadcasting Co. v. FCC, 395 U.S. 367 (1969).

105. Friends of Earth v. FCC, 449 F. 2d 1164 (D.C. Cir. 1971).

106. Columbia Broadcasting Systems, Inc. v. Democratic National Committee, 412 U.S. 94 (1973).

107. Muir v. Alabama Educational Television Commission, 688 F. 2d 1033 (5th Cir. 1982) *cert. den.*, ___U.S. ___, 75 L. ed. 2d (1984).

108. Midwest Video Corp. v. FCC, 571 F. 2d 1025 (8th Cir. 1978).

109. "Beyond the Global Village." Address on World Communications Day, United Nations, New York, May 17, 1983.

110. Which is only to say that we have sometimes been more and sometimes less successful in preserving and protecting them.

111. *Technologies of Freedom*, Belknap Press, Cambridge and London, 1983.

Comments

JORDAN J. BARUCH

Jordan Baruch Associates

What a joy it is for an engineer to have the last word after a lawyer! In this case, however, that joy is tempered by awe at Anne Branscomb's thorough and scholarly treatment of the 10 fundamental rights she associates with information. Hesitantly, I would like to add an eleventh right that also merits our concern even if it is not clearly a property right. I shall call it the right to aggregate and act upon information.

When John Mayo talked about the growing number of elements on a chip, he neglected to mention that the power of a chip is not linearly related to the number of those elements. That is an important characteristic of information as well. In information, when we have one piece of information it may be worth but little. A second piece that adds to the picture makes the total value greater, and the third piece, even greater. Indeed, the last keystone bit of information often increases the value enormously. In other words, if we were to plot the value of an information collection against the number of pieces of information in the collection we would often find a curve that grew more than linearly. In information collection, the whole is truly greater than the simple sum of the parts. Information becomes more powerful as we aggregate it, and the authors of these papers recognize that power of aggregation. Certainly Melvin Kranzberg spoke about the aggregation of knowledge in connection with the formation of Gutenberg's printing press.

I would like to point out that since the value of information is nonlinear, many of the things being discussed here get rather fuzzy and much more complicated. For example, bits and pieces of information about a commercial company may have little value taken independently; but when they are aggregated by somebody smart enough to put them together they become very important and very valuable. How do we protect the property value in the aggregate of a collection where the individual elements may each belong to others and have little value in themselves? If, in fact, we watch a smart friend use his computer in this process, what do we see? First of all, he gets his information from a large number of places through a little digital window. He knows tools to put those pieces together and draw conclusions. He builds models, examines those models, and tries to predict the near-term future. Information aggregation, model, and prediction all fuse to become a new piece of information. Its primary protection, because it is likely to be evanescent, rests with the subsequent action that it triggers. If our smart friend works well, he will prosper by acting and by protecting the value in the model whereby he acts on information aggregates.

Mel Kranzberg took the liberty of giving us Kranzberg's First Law. I would like to suggest two other laws that have a bearing on the information era because of the nonlinearity of information. First, power in a society will reside with those capable of imparting to a body of information the largest coefficient of nonlinearity—those who can take 3 things and get 12 or 17 or 93. These people—the manipulators and concluders rather than the owners—will have enormous influence. The second law, related to the first, may make Harlan Cleveland unhappy. In the first law I tried to state what would determine power in our society, namely, the ability to produce a large coefficient of nonlinearity. The second law simply says that *whatever* determines power in our society will be found to be nonuniformly distributed and therefore unfair. Some people in our society are simply going to be better at aggregating information and at drawing conclusions from it than others, and they will be considered to be taking unfair advantage of the new society just as those who were strong unfairly took advantage of neolithic society and those who had early access unfairly took advantage of colonial society. Whatever determines power will be nonuniformly distributed, and the haves will be looked at as unfair by the have-nots; I am afraid that will be as much a state of the information society as it was of the neolithic.

Information Technologies
in the Home

WALTER S. BAER

INTRODUCTION

Despite the frenetic life-styles of 1984 America, most of us spend more time at home than anywhere else. We sleep, eat, and take care of our personal needs at home. We cook, clean, and do other household chores, take care of children, pay bills, prepare taxes, and sometimes shop from home. Some of us earn part or all of our living at home. We entertain, talk on the telephone, play games, read, watch television, and otherwise relax at home. These home activities are all affected by the electronic technologies we have developed for communications, information gathering, and entertainment. Which technologies we use in the home and how we apply them to home activities are the topics of this paper.

The television set and the telephone are today the most important electronic information technologies in the home. They represent two broad categories of home information technologies that are worth distinguishing: (1) stand-alone, or one-way, receiving units, and (2) communicating, or networked, units. Besides television receivers, the first group includes radios, audio systems, microprocessors in appliances, videocassette recorders (VCRs) and players, videodisk players, videogames, cable TV converters, satellite TV receivers, hand calculators, and stand-alone personal computers (PCs). The second category includes telephone peripheral equipment such as answering machines and automatic dialers, interactive cable TV terminals, and PCs with two-way communications capabilities. The number and percentage of U.S. households with these technologies are shown in Table 1.

Stand-alone units generally show higher penetrations and faster growth than those linked to networks, for obvious technical, economic, and political reasons. It is easier to design, make, and sell new products that can be taken home and used independently than equipment that must interact compatibly with a network or with units in other locations. Moreover, networks are regulated by federal, state, or local governments, while stand-alone technologies generally are not. Consequently, the time needed for successful commercial introduction of a new stand-alone technology is measured in years, while new network technologies may require a decade or more for regulatory as well as consumer acceptance.

Nevertheless, the trend is toward interconnecting traditionally stand-alone units within the home and then linking them through communications networks to the outside world. Personal computers provide a good illustration. In 1983 only 7 percent of home computers sold were equipped with modems for external communications (Yankee Group, 1984). Only about 15 percent of the "high end" and "medium" PCs sold in 1983—PCs with memories of 64K and more—had modems. These machines, however, are rapidly displacing "low end" units in current sales (Table 2). By 1988 more than half of the home computers sold should have communications capability, allowing them to exchange messages with other PCs and to interact with external information services. The trend toward networked PCs in the home thus seems

TABLE 1   Information Technologies in the Home, 1983

Item	Number of Households With Equipment (millions)	Percentage of U.S. Households
Telephone	83	98
TV receiver	84	99
Color TV	75	89
Cable TV	33	39
Two-way cable TV	0.3	<1
Satellite TV receiver	0.4	<1
Videorecorder/player	9	11
Videogame	21	25
Home computer	8	9
With telephone hookup or modem	0.4	<1

SOURCE: A. C. Nielsen Company, Paul Kagan Associates, The Yankee Group, AT&T.

TABLE 2   U.S. Computer Sales to the Home (in thousands)

	1980	1981	1982	1983	Estimated 1984
High end	116	145	141	460	1,470
Medium	3	155	180	1,080	2,440
Low end	25	60	1,940	3,545	685
Total units	144	360	2,261	5,085	4,595
Installed base	144	504	2,765	7,850	12,446

SOURCE: The Yankee Group. 1984. Yankeevision. Boston, Mass. May:35-37. Reprinted with permission.

TABLE 3   Perceived Uses for Computers in the Home, 1983

	Percentage of Those Interested in Acquiring a Computer	Percentage of Actual Purchasers
Home budgeting/management	53	27
Education/learning	39	41
Entertainment/games	21	41
Run business from home	19	8
Programming	12	35
Do office work at home	10	8
Word processing/writing	8	10
Accessing information	7	4
Record keeping/cataloging	6	5
Home banking	5	2
Self improvement	4	2
All others	7	14
Don't know/none	9	4

SOURCE: The Yankee Group. 1984. Yankeevision. Boston, Mass. May:35-37. Reprinted with permission.

well under way. Companies such as IBM, AT&T, CBS, and Times Mirror are all betting that networks of home PCs will present large and attractive markets by the end of this decade.

How will individuals and families use their home computers? Table 3 shows the results of a recent survey of perceived applications. Home computer purchases through 1983 were predominantly "low end" machines used primarily for games and for learning about computers themselves. Purchasers are now insisting that PCs also be useful for word processing, home management, and other more sophisticated and serious applications. As a result, the number of home PCs sold in 1984 may fall below the 1983 level, but the mix has shifted dramatically toward more expensive and capable devices including larger memories, disk drives, printers, and modems.

Recent studies show growing consumer interest in and acceptance of information technologies for the home (Yankelovich, Skelly, and White, 1984). These trends in the 1980s differ markedly from the antitechnology, away-from-home attitudes of the late 1960s and 1970s. Demographics are changing as well, with significant trends towards smaller households and more working adults per household. According to the U.S. Bureau of Labor Statistics, more than half of all married women now work outside the home, up from about 30 percent in 1960. Time is thus at a premium for more Americans, and household activities must change accordingly. For middle-class U.S. households, time budget choices are at least as critical as income is in determining the purchase and use of information technologies.

In addition, more Americans seek individual, customized life-styles. We want more flexibility in our working hours and more choices among the foods we eat, the trips we plan, the magazines we read, and the television programs we watch. A central theme of this paper is that information technologies facilitate these choices but do not themselves determine them. Rather, how individuals choose to spend their time largely determines the information technologies they acquire and use. The sections that follow discuss four significant uses of time at home— working at home, doing chores at home, learning at home, and relaxing at home—and the ways in which information technologies support these activities.

Some limits on the paper's scope and intent deserve mention. It is not intended as a technological forecast, but is rather a discussion of how information technologies are being used in the home, of current trends, and of some expected developments. Technologies that will be used in the home in the next decade have already been invented and are in various stages of development, field testing, or use in business, government, and other nonresidential environments.

The paper deals with developments in the United States and assumes a business-as-usual scenario both internationally and domestically. This implies that individuals and families will continue to purchase information products and services in commercial markets. There is little discussion of equity issues or possible government programs such as subsidies for cable television and home computers.

Finally, the paper focuses on technologies for electronic communications, information distribution, and entertainment in the home. It largely neglects the interesting developments under way in the print media. It is the author's strong contention that the new electronic media will change the old without supplanting them, just as television has changed radio and newspapers without destroying the viability of either. Newspapers, magazines, books, and other forms of print media will remain viable for the foreseeable future, but ongoing changes in technology, demographics, and life-style all favor the growth of electronic information technologies in the home.

WORKING AT HOME

Personal computers provide greater opportunities to earn a living at home, especially for self-employed professionals and independent contractors such as computer programmers, consultants, writers, typists, and accountants. Individuals in these occupations generally work independently, control their own work pace, and have measurable output. Many of them could work at home without a computer, of course, but the computer's word processing, financial modeling, and record-keeping capabilities enhance their productivity. The "expert system" software now emerging from R&D will further support professionals' work at home (Hayes-Roth, 1984).

The use of space at home need not change dramatically to support such professional work. If a separate office or study is not available, most people can place a computer and related equipment in their bedrooms. The bedroom can be used for work during nonsleeping hours, with reasonable privacy and relatively little interference with other home activities.

Telework

Many employees in white-collar occupations could work at home with communications networks linking their personal computers to company data bases and to co-workers. As one illustration, a 1976 study found that the insurance industry could decentralize much of its underwriting and administrative operations by using workers at home or at dispersed sites near home (Nilles et al., 1976). Besides saving commuting costs, "telework" opens up employment opportunities to those who find it difficult to travel to work, such as parents with child care responsibilities and the physically handicapped (Olson, 1983).

Salespeople could conduct more work at home if they had ready access to customer records, as well as means to send in their orders electronically. Software programs for personal computers are available to facilitate such order entry processes. Mobile telephones using cellular radio technology also help salespeople work more effectively outside the office.

Despite these technical advances, predictions from the 1970s and early 1980s of significant shifts to working at home have not come to pass (Harkness, 1977; Nilles et al., 1976; Toffler, 1980; Williams, 1982).* Nilles now estimates that "today there are between 20,000 and 100,000 telecommuters of all sorts in the U.S." (Krier, 1984). Some forecasts simply extrapolated too far from the 1973–1974 energy crisis and concluded that skyrocketing gasoline prices would dramatically reduce commuting. Others were based on an either-or fallacy that employees would fully substitute working at home for working at an office. This is clearly not correct. Most individuals want to spend a good part of their working time among colleagues. They like the change of scene, the work-related interactions, the socializing, and even the distractions of an office environment.

However, many employees would like the freedom to work at home part of the time. Just as face-to-face contacts at the office enhance job performance, so, too, may the ability to concentrate at home, free from meetings, colleagues' visits, and business telephone calls. This paper, for example, was written largely at home, where I can find more uninterrupted time than at my office. Other employees prefer the added flexibility that part-time work at home offers for child care or other activities. Part-time work at home illustrates the customized life-styles sought by so many Americans in the 1980s.

The reporters at the *Los Angeles Times* provide an interesting example of how information technologies open new opportunities for part-time work at home. In 1982, the *Times* installed a sophisticated "front end" computer system for the editorial staff. More than 550 microcomputer terminals are linked to 21 computers in Washington, D.C., and three Southern California locations. In addition to typing and editing stories, each reporter can maintain his or her own electronic files, as well as send and receive electronic messages through the system.

The *Times'* editorial computer system was carefully designed to carry heavy peak loads, since most reporters use the system during

---

* Few quantitative data exist on the number of Americans working at home, with or without computers and communications networks. A 1984 *Futurist* article cites a U.S. Chamber of Commerce report indicating that "10 million businesses list home addresses as their place of business" (Wolfgram, 1984). A *Wall Street Journal* editorial states, "A recent AT&T study reportedly found that 11 million Americans work at home; 7 percent of the total labor force work at home full time and 6 percent pursue part time jobs at home" (Wall Street Journal, 1984). And a recent article in *Reason* cites the AT&T estimate, as well as a figure of "5-10 million, reported in *Consumer's Digest*" (Rubins, 1984).

the same hours before deadline. The editorial staff helped design the terminal configuration and functions so that the system would serve the needs of its users rather than its technical developers. Although some reporters expressed initial apprehension, most adopted the computer system eagerly within a few months and have discarded their old typewriters. This has been the general newspaper experience with editorial computer systems (Johnston, 1984).

Dial-up access was an important part of the system design so that reporters outside of Washington and Southern California could file their stories remotely. But the design did not anticipate the local reporters' enthusiasm for using the computer system from home or on assignment. Even during 1980–1982 when the system was designed, using computers for work at home was not widely accepted outside the computer field itself. Moreover, truly portable computers with built-in modems and enough capacity for serious word processing were not yet available.

All this has changed in the last two years. More than 100 *Los Angeles Times* reporters regularly use portable computers or PCs at home to write their stories and then send them electronically to the central computer system. The reporters can plan their time more flexibly, spend more time on assignment, and work closer to deadline. An increasing number of reporters now work part time at home.

The *Los Angeles Times* experience seems likely to be replicated as other businesses install computer systems for their professional employees and permit them to interact with the systems at home. Financial and accounting staff will access corporate data bases, perform analyses, and write reports from home. Managers will use electronic mail systems from home to keep in touch with their co-workers, as well as to substitute for some office correspondence and telephone calls. Professional and other white-collar workers in the academic, government, and nonprofit sectors will do the same. Information technologies make it easier for information workers to do at least some of their work at home.

Barriers to Computer-Based Telework

Significant nontechnical barriers to work at home remain, however, for those who are not self-employed. From the employer's viewpoint, there are several disadvantages:

- Part-time work at home does not save office or administrative costs, at least in the short run. The *Los Angeles Times* reporters still want their own desks in the office, each with a computer terminal, of course. In principle, an employer could reduce office space requirements if employees spent an appreciable part of their work time at home but, in practice, office space sharing appears difficult to implement.

- Work at home will cost employers more if they must pay for the home computer, communications, and other work materials. These added costs are usually easier to measure than the productivity gains attributable to work done at home. Employers' liability insurance costs will also increase if they must cover working at home.

- Many employers believe that face-to-face interactions in the office are essential, both for efficient functioning of the business and for effective use of the company's human resources. They argue that high morale, loyalty, and effective communications among employees come from working together in an office rather than from working individually at home.

- Data security problems arise when sensitive information is available at home. Processing insurance claims or conducting financial analyses requires access to confidential data. The security problems include not just wiretapping and unauthorized access to computer data bases, but, more importantly, physical security in the home environment. Homes do not have guards, locked files, or other physical protections that offices employ. Establishing these safeguards in employees' homes would impose huge costs and administrative burdens on both employers and employees. Tolerating a lower level of data security in the home may be acceptable at present, but not in the future if part-time work at home becomes widespread.

In general, management control is more difficult in multiple home locations than in a centralized office. Except at senior managerial and professional levels, working at home runs counter to established control mechanisms and policies of most corporations and government agencies.

Organized labor also opposes computer-based work at home. In the 1930s unions successfully fought against homework in the apparel industries, arguing that the minimum wage and other labor standards could not be enforced for work at home. Under the authority of the Fair Labor Standards Act of 1938, the U.S. Department of Labor in 1943 banned homework for women's garments, knitted outerwear, glove and mitten manufacturing, embroidery, handkerchief manufacturing, jewelry making, and button and buckle manufacturing. The Reagan administration's efforts to end these bans have so far been rejected by the courts (Taylor, 1983).

Unions believe that the same issues surrounding garment making will apply to computer-based homework. While agreeing that part-time work at home may be appropriate for managers and some professional employees, they contend that clerical and support workers will be paid less than minimum wage, lose benefits, and otherwise be exploited in "electronic sweatshops" (Chamot and Zalusky, 1983). The 780,000-member Service Employees International Union has banned homework by its members (Rubins, 1984). The AFL-CIO, at its Fifteenth Constitutional Convention in October 1983, formally adopted a resolution against computer homework:

> RESOLVED: That the AFL-CIO calls for an early ban on computer homework by the Department of Labor as a measure of protection for those workers entering the market for the fastest-growing occupation in the United States.

Government Policies Affecting Work at Home

Some government regulations and policies also discourage the use of information technologies for work at home. Tax policies represent a prime example. Over the last several years, the Internal Revenue Service has tightened its position on deductions for home offices and home computers. It is now very difficult for an employee to deduct such costs if the employer provides work space and equipment at the office. An employee's ability to deduct work-related costs certainly influences his or her decision about whether to work part time at home.

Zoning laws also affect the kind and amount of work that can be performed in a residential area. In Los Angeles, for example, I may legally use a computer at home to conduct business, either for myself or for my employer. However, I may not have any employees working at my home office, nor may I see any customers or suppliers at home. Chicago has a much more restrictive zoning ordinance, which prohibits professional work at home involving "installation or use of any mechanical or electrical equipment customarily incident to the practice of any such professions." The Chicago ordinance has actually been invoked to stop a teacher and his wife from writing a textbook or developing software programs on their home computer (Rubins, 1984).

Zoning laws can, of course, be changed to suit new circumstances, but only after many hearings, much effort, and a great deal of time. Like changing corporate policies and union rules, it takes longer to change zoning laws, tax regulations, and other government policies than to develop new information technologies.

In summary, technological developments, as well as changing life-styles, favor bringing more work home from the office. But the institutional policies of government, unions, and business make the shift more difficult. Businesses particularly are conservative organizations. Even though more than 100 U.S. corporations have tested the telework concept, with generally positive results, most companies still appear reluctant to change established practices to encourage its widespread application.

DOING CHORES AT HOME

Not everyone earns a living by working at home, but virtually all of us do housework. In *housework*, I include sedentary chores such as bill paying and tax preparation, as well as the physical tasks of cooking, cleaning, and home maintenance. Information technologies aid us in doing chores at home in three principal ways: (1) by controlling the appliances we use for housework, (2) by establishing local networks for home communications and control, and (3) by directly providing electronic information and transaction services from the home.

Controlling Appliances

"The technological systems that presently dominate our households were built on the assumption that a full-time housewife would be operating them . . ." (Cowan, 1983). That assumption, of course, no longer holds true in most American households. Information technology in the form of the microprocessor permits more flexible control of appliances, thus supporting the life-styles of busy people who work outside (or perhaps inside) the home.

Even as families contemplate the purchase of a home computer, microprocessors are entering their homes as control components of stand-alone appliances. Microwave ovens, refrigerators, washers, and sewing machines contain microprocessor-based controls that can proceed through a complex series of steps and respond more accurately to the surrounding environment. Ovens can be preprogrammed to defrost, brown, and heat sequentially so that dinner is ready 15 minutes after the last family member comes home from work or school. Refrigerators monitor their own operations and signal if a door is not closed tightly or a mechnical malfunction occurs.

Limits to what microprocessor-controlled appliances can do are set more by available sensors than by the microprocessors themselves. For example, it is not too farfetched to anticipate a cleaning robot in the 1990s that can vacuum automatically without stumbling into furniture. Achieving this, however, will require substantial advances and cost reductions in sensor technologies as well as processing capabilities.

The diffusion of microprocessors in the home is somewhat analogous to the home use of electric motors. Before the 1930s, electric motors were too large and expensive for most home applications. The development of cheap, fractional-horsepower motors led to the proliferation of small motorized appliances after World War II. Middle-class American households today have literally dozens of electric motors in blenders, grinders, can openers, knives, power tools, hair dryers, toys, and toothbrushes. In a similar way, inexpensive microprocessors are becoming commonplace in everyday household appliances.

Home Networks

Local networks within the home allow even more flexible control of appliances, as well as security and other home management functions. Microprocessor-based controllers use the home's electrical wiring to signal appliances from a central panel in the kitchen or bedroom. The television set can also serve as the central panel, controlled by a hand-held device similar to a remote channel switcher. With these home networks, lights or appliances can be turned on and off from another part of the house, or remotely from outside the home with a telephone call. Home networks can also help in energy management by, for example, allowing you to call home to turn on the heat in an empty house just before arriving.

"Smart telephones" and telephone peripheral equipment also support busy life-styles and aid home management functions. Thirteen percent of Los Angeles households have a telephone-answering machine, although the extent to which these devices improve the quality of life for either the caller or the answering party remains open to question. Home security systems linked to the telephone or to cable TV networks also are in demand. AT&T now offers a microprocessor-based smoke alarm connected to the telephone that, when triggered by smoke, dials a preprogrammed number and announces, "Fire! at [the specified address]." Developers commonly offer microprocessor-controlled intrusion alarm systems as amenities in newly constructed apartments, condominiums, and houses.

Home Information and Transaction Services

Personal computers and home terminals linked to external data bases will encourage more electronic information gathering and transactions from the home. The telephone, of course, has provided such services for more than a century. Although similar computer-based services for banking, shopping, ticket ordering, and information gathering have been relatively slow to develop, several now appear near the takeoff point:

- Electronic banking at home is available from more than 30 financial institutions. Bank of America and Chemical Bank together have more than 20,000 electronic home banking customers. Bank customers use these electronic services to write checks, review account balances, compare current interest rates, and move funds among their accounts, as needed. Customers particularly like the ability to review their account balances at any time, without waiting for a monthly statement to arrive in the mail.
- Securities transactions and airline reservations now can be made electronically from home PCs via CompuServe and The Source, two popular computer information services.
- Videotex offers better color, graphics, and user-friendly interfaces for home transaction and information services. Consumer videotex services with high-quality graphics are now operating in Miami, Florida, and Orange County, California.* The Times Mirror Gateway service in Orange County offers its subscribers electronic home banking, financial planning, ticket ordering for entertainment and sports events, travel booking, catalog ordering, and home shopping from department stores and other retailers.
- Stock and commodity price quotations are available at home from Dow Jones, Commodity News Service, and several other vendors. Most customers of these financial information services use the two-way telephone network to transmit queries from their home computer to a central data base, which then sends the requested information back to the home. If two-way transactions are not required, however, the information alone can be broadcast as a one-way service. Individual stock and commodity quotations are currently available via FM radio subcarrier and direct broadcast satellite. These services are now priced so high that they appeal only to active investors, but their costs should drop within the next few years to make them accessible to more households.
- Financial management draws considerable consumer interest (Table 3), although few individuals as yet use their home PCs very effectively for budgeting, tax preparation, or other financial planning tasks. As more and more people become familiar with spreadsheet software programs in business, their use in the home for financial management will undoubtedly increase.

These home transaction and information services are still in their infancy. Fewer than 1 percent of the nation's households subscribe to

---

* These videotex services use the North American Presentation Level Protocol Standard (NAPLPS), which provides more colors and higher-resolution graphics than alphamosaic services such as Prestel in the United Kingdom, or text-only, ASCII services such as Dow Jones News Retrieval, CompuServe, or The Source.

any of them. Yet consumer attitudes toward electronic transactions from the home have turned decidedly more positive within the past year (Table 4). Acceptance has moved beyond the small number of early adoptors, who are the first to try anything new, toward a broader segment of the public. From the field trials and commercial experience to date, both information and transaction services appear necessary for the commercial success of videotex and similar electronic services to the home. Neither pure information nor pure transactions seem viable without the other.

Home transactions appeal especially to adults working outside the home, for whom time is at a premium. Consumer surveys show that busy adults highly value information that saves them time or money (Yankelovich, Skelly, and White, 1984). Consumers also prefer the flexibility of electronic transaction services that are available 24 hours a day, 7 days a week. They do not want to depend on traditional banking or shopping hours.

With electronic transaction services at home, consumers will do more of the work themselves and rely less on middlemen such as bank tellers, order clerks, and travel agents. Banks' increasing use of automatic teller machines (ATMs) already portends this trend. The next step will be to move from the ATM to a home terminal for all but cash transactions. As a result, employment opportunities for tellers, clerks, and other service agents may be squeezed as home transaction services expand.

Electronic transaction services in the home will not replace retail stores or service businesses, however. As they do when earning a living, people like to mix with other people and are not likely to avoid

TABLE 4 Consumer Perceptions of Information Technologies for Transactions at Home, 1983 and 1984

	Percent Responding	
	1983	1984
Computers that allow for at-home purchasing		
—Make life better	16	26
—Make life worse	23	20
Bank at home via personal computer or cable TV		
—Makes life better	21	30
—Makes life worse	20	18

SOURCE: Yankelovich, Skelly, and White, Inc. 1984. The Yankelovich Monitor. New York. Reprinted with permission.

all shopping trips. Their buying habits may change, though, with more electronic ordering of commodity products and services done from home.

LEARNING AT HOME

Parents express great interest in using personal computers to teach their children at home. As indicated in Table 3, more than 40 percent of the home computer buyers surveyed expect to use them for learning and education. Computer manufacturers know these statistics, too. Their advertising emphasizes the importance of computer literacy and exhorts parents to prepare their children for success in the classroom and the job market. Software packages clearly labeled as educational stay on the best-seller lists at home computer stores. Educational software accounted for about $100 million, or just under 10 percent of all software sales for home computers in 1983 (Holden, 1984).

It is important, however, to distinguish between learning and formal education. Computers and other information technologies unquestionably can aid learning at home, especially when the educational content is entertaining. Flight simulation programs, for example, effectively teach elements of aviation even though they are designed as games. Some educational software programs are fun to use as well, for example, "Rocky's Boots," which introduces concepts of computer logic, and "Green Globs," a game that teaches analytic geometry. But many more programs are pedestrian and dull. Like earlier examples of educational technology, they will be put on the shelf and soon forgotten.

Personal computers are not the only information technology for learning at home. Videotapes such as "Jane Fonda's Workout" have considerable learning potential. So do conventional broadcast television programs, as "Sesame Street," "Mr. Rogers," the National Geographic specials, and Bill Moyers' "A Walk Through the Twentieth Century" attest. Research confirms the layman's perception that television is a particularly powerful medium for learning. As Patricia Greenfield reports in her recent review of media effects: "Children tend to learn what they see on television more thoroughly than what they read or hear on radio or tape" (Greenfield, 1984).

Whether computers will move the locus of education toward the home is a far different question. Several factors militate against the change:

- We have heavy investments in schools' physical plant and in teachers' human capital. The educational establishment is unlikely to show much enthusiasm for moving formal learning from the schools into the home, even if technologies permit or favor the change.

- Few parents want to transfer education from the school to the home. Working parents count on the schools to take charge of their children during the day. If anything, adult life-style changes imply heavier reliance on the schools for activities once carried out at home.
- Children, like adults, want face-to-face contacts with their peers. Schools facilitate these relationships and the learning that accompanies them.
- We remain in large part a credential society, demanding certificates from schools as prerequisites for jobs and advanced education. Although home-based degree programs exist, they generally do not substitute for the credentials given by the schools.

Still, America's current disenchantment with the schools and the calls for educational reform undoubtedly will stimulate more parents to buy computers and educational software for learning at home. The educational consequences are likely to be mixed. One probable result will be to widen the knowledge gap between children from affluent and lower-income homes. According to a recent review of a National Institute of Education conference on the topic, "the vast majority of computer buyers are well off, well educated and white. Sixty percent of those purchasing computers costing over $500 have incomes over $40,000" (Holden, 1984). In addition to learning with their computers at home, children from these affluent families will be more comfortable using computers at school and eventually on the job.

The home computer should not be branded the villain, however. This same gap-widening problem has occurred for every important new learning tool, including paperback books, telephones, and television. Each of these developments began by helping the affluent more than the nonaffluent. Although "Sesame Street" was created principally to teach verbal and number skills to children from low-income families, which it did very successfully, children from affluent homes ended up learning even more (Cook et al., 1975). Computers in the home simply add another dimension to the knowledge gap between children from different socioeconomic backgrounds.

One partial solution to the computer learning gap at home is to provide computer instruction to all children at school. Although more than two-thirds of U.S. public schools have at least one microcomputer, they still are not widely available for student use (Market Data Retrieval, 1984; Shavelson and Winkler, 1984). A 1983 study, cited by Shavelson and Winkler, found that the ratio of students per microcomputer averaged 183 to 1 in elementary schools, 181 to 1 in junior high schools, and 125 to 1 in public high schools. Parental purchases of home computers thus may place increased pressure on the schools to add more computers in the classroom.

Opportunities also may exist to coordinate home and school use of microcomputers. Word processing provides a good illustration. As Seymour Papert pointed out in his celebrated book *Mindstorms*, computers enable children to improve their writing markedly through editing and reworking their thoughts on the screen:

> For most children rewriting a text is so laborious that the first draft is the final copy, and the skill of rereading with a critical eye is never acquired. This changes dramatically when children have access to computers capable of manipulating text. (Papert, 1980)

Children can be introduced to word processing at school and then reinforce and extend their writing skills on a home computer. Coordination would leverage the educational impact of both home and school computing.

Unfortunately, there seems little precedent for effective coordination between school and home in the use of other information technologies such as television. The conceptual foundations exist for involving parents in teaching critical television viewing skills to children (Dorr et al., 1980) or for incorporating prime-time television programs into the school curriculum, but little implementation has occurred. Perhaps computers in the home will be different from other educational technologies and will offer better opportunities for home-school coordination.

RELAXING AT HOME

Home is the principal place for leisure-time activities, including entertaining, talking on the telephone, reading, and watching television. Time budget studies show that television viewing dominates the use of free time at home (Table 5). Adults in 1975 watched television as a primary activity 2.2 hours a day, on average, a substantial increase from the 1.5 hours recorded in 1965 (Robinson, 1977, 1981). More recent data show a slight decline in adult viewing since 1975, but it still averages more than 2 hours a day (Juster, forthcoming).

These data fall below other figures cited by the Television Information Office and the rating services (A. C. Nielsen, 1984; Roper, 1984). According to recent Nielsen data, home television sets are turned on an average of more than 7 hours a day. The differences result from two factors: different household members watch the same television set at different times during the day; and individuals often watch television as a secondary activity during time primarily devoted to

TABLE 5   Time Spent in Leisure Activities, 1965 and 1975

Activity	Minutes per Day	
	1965	1975
Study	15	23
Organizations	16	16
Entertainment	21	17
Visiting (and meals out)	64	54
Active leisure	18	22
Leisure travel	21	24
Television	89	132
Reading	35	31
Other	24	24
Total	303	343

SOURCE: Adapted from "Television and Leisure Time: A New Scenario," by John P. Robinson in the Journal of Communication, Vol. 31, No. 1.

eating, personal care, housework, reading, or other leisure activities. The Nielsen data show average daily TV usage continuously increasing from 5.5 hours in 1965–1966 to 7.5 hours in November 1983. Households with three or more people, as well as those with pay cable services, watch more than 8.5 hours per day (Figure 1). Television viewing declines with increasing household income, although the difference between high- and low-income households is relatively small.

The growth of television viewing has clearly affected other activities in the home. Since 1950, television has taken time away from sleeping, housework, reading, and conversations with friends (Comstock, 1982). President Reagan, someone quite familiar with the medium, observes that television substitutes for face-to-face visits:

> We watch a lot of television, seeking continuity and reassurance in the regular and predictable appearance of our favorite TV stars and programs. They visit us—as if they were a friend or relative coming by for the evening. TV is increasingly becoming the American neighbor. (Reagan, 1984)

Just as the automobile transformed the ways in which individuals and families deal with distance, so television has profoundly changed their use of time. We still do not understand the full personal and social implications of this shift to television viewing in the past generation.

Technical Improvements in Television

Given the vast market that television represents, much technical and economic activity is directed toward expanding and improving it.

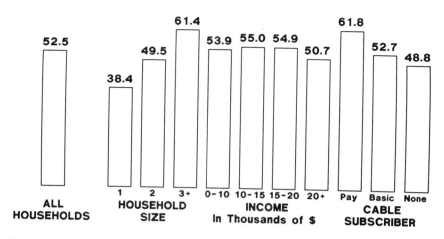

FIGURE 1   Hours of TV usage per week by household characteristics, November 1983. SOURCE: A. C. Nielsen Company. 1984. Nielsen Report on Television. Northbrook, Ill. Reprinted with permission.

Twenty years ago, only 8 percent of U.S. households could receive nine or more television stations, while today 61 percent can do so (A. C. Nielsen, 1984). Cable television brings even more viewing options, including 24-hour-a-day channels devoted to movies, sports, news, weather reports, government events, and contemporary music. Many cable systems built in the past five years have the technical capacity to deliver 50 or more TV channels to the home, but not yet enough programming to fill them.

Continuing technical development, along with deregulation of the electromagnetic spectrum available for television, will make more channels of television available to U.S. households via several distribution means:

- *Low-power television* (LPTV) uses directional antennas to broadcast additional television signals at conventional VHF and UHF frequencies without interfering with existing stations. Although few LPTV stations are yet on the air, the Federal Communications Commission (FCC) has awarded more than 200 licenses by lottery and continues to process the several thousand applications it has received.

- *Multichannel, multipoint distribution service* (MMDS) broadcasts at microwave frequencies previously reserved for instructional television. MMDS services can bring at least eight new television channels to homes in metropolitan areas. The FCC is also holding lotteries to award MMDS licenses.

- *Direct broadcast satellite* (DBS) could beam 12 or more television channels to the home by the end of this decade. Although some early entrants have withdrawn from the field, DBS service still holds promise, especially for rural households that are not well served by conventional broadcast and cable television.
- *Fiber optic systems* may in the 1990s overtake coaxial cable for video distribution to the home. The wired video distribution networks of the twenty-first century probably will be switched, digital, and fiber optic. Both the telephone and cable television industries are positioning themselves to provide fiber optic systems in the future (Baer, 1984).

Efforts are also under way to improve the image and sound quality of television receivers:

- *Stereo sound reception* is already available in top-of-the-line receivers. Only a few TV stations are as yet broadcasting in stereo, but the industry expects consumer demand to accelerate and to prove a boon to TV set manufacturers. Some TV programs will carry bilingual sound tracks instead of stereo.
- *Digital TV sets* contain a few digital chips that replace several hundred analog components. They might be more accurately termed "digitizing" TV sets, since they convert an analog TV signal to digital, process it digitally, and then reconvert it back to analog for display. Digital sets improve image and sound quality by simplifying TV signal alignment, reducing ghosts and noise, and compensating for aging components. The built-in digital circuitry should lower the cost of teletext and videotex decoding, as well as provide new features such as freeze-frame and split-screen viewing. Already available in Europe, digital TV sets will be on the market in the United States by early 1985. Many industry experts expect digital sets to capture the bulk of the television receiver market within a few years.
- *Component video systems* separate the television display monitor and audio speakers from the tuner and other picture and sound processing components. Component systems appeal primarily to serious videophiles who want better image and sound quality, particularly when playing videotapes and videodisks.
- *High-definition TV* (HDTV) would replace the 525-line NTSC standard, which has prevailed in North America since the 1940s, with images containing 1,000 or more lines. At this resolution, the perceived image quality is about the same as that of 35mm film. In addition to more lines, HDTV also will have a wider aspect ratio, multichannel sound, and a better technical method of transmitting the brightness, color, and sound information within the available TV bandwidth. Technical groups are now studying the issues surrounding HDTV and hope to reach agreement on international standards for HDTV systems. Digital signal processing within the TV receiver can help deal with compatibility problems between HDTV

and existing television standards. However, because digital signal processing can clean up lower-resolution images from existing standards and make them look better to viewers, digital TV sets may actually hold back the demand for HDTV.

• *Projection TV* and other large-screen displays are improving in quality but remain limited in consumer appeal by the low resolution of NTSC television images. Digital signal processing or high-definition TV will make large-screen display much more attractive to consumers. Work on flat panel displays is also progressing, although these new technologies still find it difficult to compete in cost with the venerable television CRT.

Consumer purchases of videocassette recorders (VCRs) have surpassed even optimistic projections of a year ago. The installed base of home VCRs has grown from less than 1 million in 1979 to 9 million by the end of 1983. Sales in 1984 are up over 60 percent above 1983, with no signs of leveling off. The industry estimates that 16 percent of U.S. households have VCRs now, and projections are for 30 percent market penetration by 1986, and perhaps 50 percent by 1990. Consumers want VCRs both to record programming from broadcast or cable TV for viewing at more convenient times and to play prerecorded tapes. The prerecorded tape rental business has grown as rapidly as VCR sales, giving VCR owners an even wider range of program choices. The impact of VCRs on home activities is discussed below.

Videodisk players may make a comeback in the home, despite RCA's well-publicized exit from the business after losses of nearly $600 million. Optical videodisk players offer better image and sound quality than VCRs or the CED videodisks that RCA marketed. The anticipated success of "compact" audiodisks using similar technology for high-quality sound reproduction may reawaken interest in optical videodisks as well. So, too, may the use of the optical videodisk's interactive capabilities for games and, eventually, information retrieval. Following their success in video arcades, several companies are developing videodisk games for the home. Still, videodisk players are likely to have a narrow appeal to affluent early adopters rather than appealing to the mass home market.

These technical developments in distribution systems, television receivers, VCRs, and videodisks continue the trend toward making more video channels available in the home with technically better image and sound quality. The number and variety of available video programs also are increasing, although not as rapidly as the channels for distributing them. Opinions differ widely as to whether the *quality* of video programming available in the home has improved. Survey data suggest that consumers perceive the new video technologies more positively than conventional television programming, which more respondents think has worsened rather than improved in the past year (Table 6). However, the ratio of "worsened" to "improved" responses has changed from nearly 3 to 1 in 1980 to about 3 to 2 in 1984. By comparison, the perceived change in cable/pay TV quality in 1984 is positive by more than a 3-to-1 ratio. Consumer attitudes toward videotape recorders have remained much more positive than negative over the past four years.

New Technologies' Challenge to Television

Other information technologies seek to challenge television for the individual's leisure time. Videogames have been the principal diversion, especially for children. Videogames have been purchased by one-quarter of American households, or nearly 40 percent of all households with children. But enthusiasm for action videogames is declining, at least among adults who respond to surveys (Table 6). More people now think these games "make life worse" than "make life better," a reversal of attitude in the past two years.

Interest is turning away from videogame-only consoles to more sophisticated games played on home computers. They range from software versions of Trivial Pursuit to board games (Scrabble, backgammon, chess) to interactive fiction in which players create their own

TABLE 6 Consumer Perceptions of Home Entertainment Technologies, 1981–1984

| | Percent Responding | | | |
	1981	1982	1983	1984
Quality of TV programming				
—Improved	19	16	22	27
—Worsened	55	54	48	42
Quality of cable/pay TV				
—Improved	—	—	—	47
—Worsened	—	—	—	15
Videotape recorders				
—Make life better	25	22	19	26
—Make life worse	9	11	10	10
TV games				
—Make life better	24	22	19	20
—Make life worse	12	17	23	24

SOURCE: Yankelovich, Skelly, and White, Inc. 1984. The Yankelovich Monitor. New York. Reprinted with permission.

characters and participate in an adventure story (Zork, Witness). A recent book on the psychological implications of computers describes how compelling interactive fiction games can be for children and teenagers (Turkle, 1984)—and, I suspect, for adults.

Up to now, electronic games in the home have been stand-alone units connected to the television set or to the computer display. Downloading games via telephone, cable, or satellite is technically feasible and may prove economically attractive to consumers. Several companies are now testing the commercial prospects for downloaded games. Games are also a prominent feature of videotex and other interactive home information services. Videotex games are now played interactively between a player at home and the host computer. However, technical advances in microprocessor hardware, software, and communications are proceeding to a point where multiplayer interactive games may be offered commercially within the next two or three years. Such games would link players in separate households via cable television or the telephone network, with the host computer acting as referee. As Robert Lucky of AT&T Bell Laboratories has suggested, videotex and communications companies' new advertising theme may be "Reach out and play someone" (Lucky, 1984).

Videotex brings information on demand, including (for leisure activities) news, sports, and other features. It also enables subscribers to send messages and chat electronically. Electronic bulletin boards on such topics as movie reviews, recipes, and dating have proved very popular in videotex and other home information services around the country. Videotex will not replace the telephone for person-to-person communications or supplant newspapers and other print media for most information needs, but it seems likely to find a niche among media uses in the home.

Will any of these new, more interactive services displace television viewing? If they are to have significant impact on leisure activities, they must take some time away from television. There is simply very little time in other home leisure categories to replace.

Social commentators often criticize television viewing as a passive, lowest-common-denominator activity. Despite more program options and better technical image quality, most intellectuals still consider television a "vast wasteland." They would like to see television viewing replaced by more active pursuits, such as learning, conversing with others, and even game playing. But the critics' concerns are not really relevant here; individual preferences are the real issue.

Prospects for displacing television with other activities are supported by studies showing that television provides relatively low consumer satisfaction. Robinson reports the results of a 1975 survey in which respondents rated various activities on a scale from 0 (dislike a great deal) to 10 (like a great deal):

> While the average score of 6.1 for TV viewing did fall on the positive side of the scale, it was well below the scores of almost all other free time activities. . . . Television was rated about the same as housework for women (but not for men) and well below work. (Robinson, 1981)

Opinion surveys conducted over three decades show growing consumer dissatisfaction with television. Half of a sample surveyed in 1980 agreed that "television seems to be getting worse all the time," compared to 24 percent in 1960 (Bower, forthcoming). These results are consistent with the Yankelovich findings shown in Table 6. More consumers consistently say that the quality of television programming has "worsened" than that it has "improved."

Other media experts contend, however, that television precisely meets the needs of tired individuals who seek relaxation at home. They want passive, undemanding entertainment during their evening leisure hours. Watching television is habitual, Russell Neuman concludes after reviewing studies of viewer behavior:

> The viewer plops down in front of the set, spins the dial, examines the programs available, and selects the least objectionable. Surveys repeatedly confirm that most viewers report "watching whatever is on.". . . Close analysis of viewing behavior indicates there is almost no correlation between expressed preferences and actual viewing behavior. (Neuman, forthcoming)

Television may generate a low level of viewer enthusiasm, but it has a high level of acceptance.

Videogames may be displacing some television viewing time among children, although the data are still quite sparse (Greenfield, 1984). In one study reviewed by Greenfield, children watched less television after they received home videogames. Whether this change will persist after the novelty of the videogame wears off is not really known.

With the possible exception of videogames, there is no convincing evidence as yet about the mass entertainment appeal of other information technologies and services. Multiplayer interactive games have largely been confined to computer hackers and students with access to powerful mainframes. Videotex may over time prove popular for leisure-time activities, as well as for transactions and information gathering, but videotex will not be a mass market service for several years to come. In general, the most enthusiastic users of interactive services are likely to be infrequent users of television. Consequently, the burden of proof still falls on those new activities that would displace the time individuals now spend watching television.

A separate but related question is the significance of the shift toward watching videotapes rather than television. Consumers use their VCRs to shift viewing times to suit their convenience. Time shifting of daytime soap operas for evening viewing is a frequently cited example. Some viewers even prefer to record sports events for later replay. They must then avoid listening to the results on the radio or reading them in a newspaper before watching their tape. Viewers also like the choice of content available from tape rental stores. Tape rentals further fragment the viewing audience and contribute to the trend toward watching less network programming.

Television advertisers already are concerned about VCR viewers who fast-forward through commercial breaks. Such "zapping" is evidently widespread. An A. C. Nielsen survey last fall found that 65 percent of VCR owners use their fast-forward button to zap commercials.* Although one cannot extrapolate too far from these data from relatively affluent households, zapping cuts to the heart of advertiser-supported television in the United States. The impact could be great as VCR ownership moves toward 50 percent of U.S. households.

Videotape and videodisk players also permit a form of "browsing" through video materials that cannot be done with conventional television. Viewers can speed up or slow down a sequence, stop to watch a still frame, and go back to take another look at a sequence of particular interest. Video browsing seems confined to a small number of serious videophiles today, but it could become a routine way to watch video programming when VCR ownership becomes widespread.

THE IMPACT OF INFORMATION TECHNOLOGIES ON HOME ACTIVITIES

How information technologies will affect these four broad categories of home activities—working, doing chores, learning, and relaxing—is

---

* I recall one recent evening when a TV viewer pressed the fast-forward button on his VCR controller as a commercial appeared, and nothing happened! "You mean we're watching in real time?" he said with impatience and incredulity.

summarized in Table 7. Up to now, technology has had a modest influence on working at home. That may change, since computer and communications technologies make it possible for more people to substitute working at home for working in an office. For most workers, however, the substitution will be partial rather than complete. And it will be slow. Information technologies have the potential to transform work at home, but decisions by individuals and institutions seem likely to bring about far more modest changes in the next decade.

Households make relatively modest use of information technologies—principally the telephone—for doing chores. However, the trend seems clear for significantly increased use of microprocessor controls, home networks, and electronic transaction services over the next decade. Yet their long-term impact on household productivity or on the perceived quality of life may be quite modest. These technologies primarily support the life-styles of busy people seeking to minimize the time and toil of household chores. They do not change the nature of these chores materially, nor do they elevate them to more satisfying pursuits. Rather, just as electricity and the telephone moved from being household luxuries to being necessities, so these newer information technologies will gradually become integrated into routine home management tasks.

Technology's impact on learning at home is much less clear. Past experience with other educational technologies should make us skeptical about extravagant claims for the computer. But computers just might be different. Some of the recent studies of kids and computers hint that, over the long term, computers might really bring about profound changes in learning. The field is moving so rapidly that we have little notion of what state-of-the-art learning software will be even a few years from now, in either the school or the home environment.

Television already has significantly influenced leisure time at home. The new technologies—cable, direct broadcast satellite, VCRs, interactive games, and videotex—will not displace conventional TV. Nevertheless, they seem likely to fragment further the mass audience for

TABLE 7   Impact of Information Technologies on Home Activities

Home Activity	Present Impact	Likely Impact in Next Decade	Potential Long-Term Impact
Working	modest	modest	significant
Doing chores	modest	significant	modest
Learning	modest	?	significant
Relaxing	significant	significant	significant

television and otherwise significantly affect leisure-time habits for a substantial number of people.

These specific activities in the home raise broader questions of how information technologies may affect personal relationships, family structures, and the overall concept of the home itself. Certainly information technologies give individuals more options as to what activities they can perform at home:

- Earning a living
- Learning
- Communicating with others
- Playing and relaxing

They encourage our efforts to customize our activities and our lifestyles.

Information technologies also expand our links with other people beyond the physical limitations of home, city, and even nation. Scientists and engineers well recognize that our communities of interest extend throughout the United States and into other countries. The telephone has been the principal technology permitting us to build networks of colleagues and friends, largely independent of geography. We have established what Melvin Webber called "communities without propinquity." Computer networks available at home as well as in the office will enlarge the range and scope of these relationships, enabling us to share common interests and experience.

But technology, a neutral force, could also serve to isolate rather than to unite us. Max Frisch explains: "Technology is the knack for organizing the world so we don't have to experience it." As is often the case, novelists have seen these issues long before technologists. Seventy-five years ago E. M. Forster gave us an apocalyptic vision of information technologies in the home in his short story "The Machine Stops":

> Imagine, if you can, a small room, hexagonal in shape like the cell of a bee. It is lighted neither by window nor by lamp, yet it is filled with a soft radiance. . . . An armchair is in the center, by its side, a reading desk—that is all the furniture. And in the armchair there sits a swaddled lump of flesh—a woman, about five feet high, with a face as white as a fungus. It is to her that the little room belongs.
>
> There were buttons and switches everywhere—buttons to call for food, for music, for clothing. There was the hot bath button, by pressure of which a basin of (imitation) marble rose out of the floor filled to the brim with a warm deodorized liquid. There was the cold bath button. There was a button that produced literature. And there were, of course, the buttons by which

she communicated with her friends. The room, although it contained nothing, was in touch with all that she cared for in the world. (Forster, 1968)

This is the nightmare of information technologies used not to enhance experience, but to avoid it. Technology *could* be used to create wholly artificial environments and to substitute for human interaction. Technology could reduce the electronic cottage to the electronic isolation booth.

Forster's nightmare stands diametrically opposed to Orwell's nightmare, with which we are more familiar, of information technologies used by a malevolent government to spy on individuals in their homes. Although it might be technically feasible for Big Brother to watch us at home via two-way television, we have not chosen that direction. The Orwellian nightmare of information technology used to entrench despotic authority and stamp out individual freedom does not seem a likely possibility today in the United States or other democracies. We recognize that technology need not determine our destiny.

Perhaps we should pay more attention to the Forsterian nightmare of electronic isolation, although it, too, does not seem a likely prospect. Humans are still largely social beings. Despite the solipsistic possibilities offered by television, videogames, and electronic shopping at home, people like to be with people much of the time. Americans believe that technology makes it easier, not more difficult, to maintain close, personal contact with other people (Yankelovich, Skelly, and White, 1984). Nevertheless, we must remain aware of the possibilities for misuse of information technologies as we seek to develop and commercialize them.

The key feature of information technologies in the home is that they give more control to individuals. They permit us greater freedom to control where and when we work, do chores, learn, play actively, and are entertained. They offer us more choices of materials in more convenient forms, for entertainment and learning. They provide us with more flexibility in obtaining information, making transactions, and communicating with individuals and institutions. They may even over time replace some passive television viewing with more active pursuits.

Of course, how these technologies will be used in the home and how the home itself evolves depend far more on nontechnical than technical factors. Technology largely supports ongoing changes in individual life-style choices, family structure, and the relationships among the home, the school, the office, and other societal institutions. It is probably not correct to say that information technologies themselves will transform the home environment. They are more responsive than causal. They do, however, support distributed decision making more than centralized authority, and they encourage some transfer of control from entertainment producers to viewers, from bosses to workers, and from instructors to learners. By empowering individuals' decisions at home, information technologies can enhance the functions of the home in an information-rich society.

REFERENCES

A. C. Nielsen Company. 1984. Nielsen Report on Television. Northbrook, Ill.

Baer, Walter S. 1984. Telephone and cable companies: partners or rivals for video distribution? In Competition of Video Media, Eli M. Noam, ed. New York: Columbia University Press.

Bower, Robert T. (Forthcoming) The Changing Television Audience in America. New York: Columbia University Press.

Chamot, D., and J. L. Zalusky. The Electronic Sweatshop: The Use and Misuse of Work Stations in the Home. Paper presented at National Executive Forum, National Academy of Sciences, Washington D.C., November 9-10, 1983.

Comstock, G. 1982. Television and American social institutions. Pp. 334-348 in Television and Behavior: Ten Years of Scientific Progress and Implications for the Eighties, Vol. 2. Rockville, Md.: U.S. Department of Health and Human Services.

Cook, T. D., H. Appleton, R. F. Connor, A. Shaffer, G. Tomkin, and S. J. Weber. 1975. Sesame Street Revisited. New York: Russell Sage.

Cowan, R. S. 1983. More Work for Mother. New York: Basic Books.

Dorr, A., S. B. Graves, and E. Phelps. 1980. Television literacy for young children. Journal of Communication. Vol. 30, p. 71-83.

Forster, E. M. 1968. The machine stops. Collected Tales of E. M. Forster. New York: The Modern Library. First published in Oxford and Cambridge Review, 1909.

Greenfield, P. M. 1984. Mind and Media. Cambridge, Mass.: Harvard University Press.

Harkness, R. C. 1977. Technology Assessment of Telecommunication/Transportation Interactions. Menlo Park, Calif.: Stanford Research Institute.

Hayes-Roth, F. 1984. The machine as partner of the new professional. IEEE Spectrum, June:28-31.

Holden, C. 1984. Will home computers transform schools? Science, July 20:296.

Johnston, D. 1984. A newsman's view of technology. IEEE Spectrum, June:112-113.

Juster, F. T. (Forthcoming) A note on recent changes in time use. In Time, Goods and Well Being, F. T. Juster and F. P. Stafford, eds. Ann Arbor, Michigan: University of Michigan, Survey Research Center.

Krier, B. A. 1984. Futurists compute the influence of computing. Los Angeles Times, September 21:V1, V24.

Lucky, R. 1984. "Entertain me": television's omnipresence. IEEE Spectrum, June:85-89.

Market Data Retrieval. 1984. Microcomputers in schools, 1983-84: a comprehensive survey and analysis. Education Week, September 5:L45.

Neuman, W. R. (Forthcoming) The media habit. In Electronic Publishing Plus, Martin Greenberger, ed. White Plains, N.Y.: Knowledge Industry Publications, Inc.

Nilles, J. M., F. R. Carlson, P. Gray, and G. J. Hanneman. 1976. The Telecommunications-Transportation Tradeoff: Options for Tomorrow. New York: Wiley Interscience.

Olson, M. H. 1983. Remote office work: changing work patterns in space and time. Communications of the ACM, March:182-187.

Papert, S. 1980. Mindstorms: Children, Computers, and Powerful Ideas. New York: Basic Books.

Reagan, R. 1984. Remarks at the Presentation Ceremony, 1983 Young American Medals for Bravery. Presidential Document. Administration of Ronald Reagan, August 28.

Robinson, J. P. 1977. How Americans Use Time. New York: Praeger.

Robinson, J. P. 1981. Television and leisure time: a new scenario. Journal of Communication, Winter:120-130.

Roper, B. W. 1984. Trends and Attitudes Toward Television and Other Media: A Twenty-Four Year Review. New York: Television Information Office.

Rubins, D. 1984. Will the plug be pulled? Reason, October:24-32.

Shavelson, R. J., and J. D. Winkler. 1984. Realizing the potential of information technology in American education. MITI Journal, October.

Taylor, S., Jr. 1983. Court reinstates federal rules affecting industrial home work. New York Times, November 30:A25.

Toffler, A. 1980. The Third Wave. New York: William Morrow.

Turkle, S. 1984. The Second Self. New York: Simon & Schuster.

Wall Street Journal. 1984. No work place like home. February 23:23.

Williams, F. 1982. The Communications Revolution. Beverly Hills, Calif.: Sage Publications.

Wolfgram, T. H. 1984. Working at home. The Futurist, June:31-34.

Yankee Group. 1984. Yankeevision. Boston, Massachusetts, May:35-37.

Yankelovich, Skelly, and White, Inc. 1984. The Yankelovich Monitor. New York.

Comments

ROLAND W. SCHMITT
Senior Vice-President, Corporate Research and Development
General Electric Company

I think Walter Baer's paper is an excellent overview and delineation of this field of information technologies in the home. It codifies and defines the terms of reference for discussion of this topic probably better than any I have seen in the past. The issues he raises have two dimensions. He has highlighted one of these dimensions—that of working, doing chores, learning, relaxing. There is also that other axis of issues that he discussed—the question of how certain is the likely future? When is it going to happen? What surprises are going to be there? I want to dwell on a couple of topics pertaining to this second axis. First, however, let me say that one of the key observations that Dr. Baer has made is that when you are looking at information technologies in the home, the key parameter is not discretionary dollars, but discretionary minutes and discretionary hours. I think that is something you must keep in mind when talking about this topic.

I would like to deal with two issues. First, markets. You might say it is a little dangerous for a technical person to deal with markets, but frankly, it is my experience that the marketing people do such a poor job of dealing with high-technology markets that I do not fear to tread on that ground. Second, I will make a few comments on what is known as user-friendliness.

Home information technology markets have some interesting characteristics. First, the user and the buyer of these technologies are really very closely linked. Most often they are the same person. That is a little different from the situation in a factory or office environment. A great number of noneconomic, nonfunctional considerations go into the reaction of the consumer; things like taste, fashion, curiosity—many of the factors that Walter Baer commented on—are very important. Therefore, I believe that these markets are really very hard to predict, and they are especially hard to mastermind. There is a lot of buying one and buying it once—the Hula-Hoop example, if you like. Consider the evolution of videogames. The first ones really did not impress people very much, and most people dismissed them as a minor novelty. Suddenly, videogames came of age, and all at once we were hearing sweeping statements about a Pacman culture. We heard that videogames were the prototype of the industry of the future. You may remember that phrase, not heard very often anymore, Atari democrat. The pop sociologists told us that the videogame was either saving or ruining our youth, depending on whom you listened to. I recall a widely publicized seminar, held at Harvard University, on the educational impact of videogames. Remember how towns were slapping curfews on video arcades? Look at it today. The arcades are going broke. Those same prophets who were wrong on the two previous rounds are confidently predicting that the videogame is dead, and I suspect that they are just as wrong this time.

Another example is the videocassette recorder (VCR). There was a very long period of latency before the VCR really hit its stride and became the major industry it is today. One of the big factors in that growth was the emergence of a mom-and-pop industry, the videotape leasing and renting industry. Suppose, for a minute, that someone in a large corporation had been smart enough early on to have envisaged the whole synergistic system that would be required to make the VCR industry boom. Suppose they had seen the necessity of establishing a network of video rental outlets and had put the ideas and scheduling together in a business plan. The answer from the top would have been that the investment is too high and the payback is too long. It just would not have gone. The present boom and growth has truly emerged, in my opinion, from the patient offering of a new technology, from experimenting with it out there in the real market, depending on the vision, on the interest, on the entrepreneurship of a lot of people who are trying things, finding out what happens, and finding out what combinations will work. Given this background, I think we have to realize that as we go down the path of evolving information technologies in the home, it is going to be a highly experimental path, one of trial and error, and one where there are going to be a lot of contributions from a lot of different people. Therefore, again, I do not think that these markets will be easily masterminded.

I would now like to turn to the topic of user-friendliness. It is a phrase we hear time and time again, and to which we attach a lot of emphasis and importance. Frankly, I think we are doing a poor job of it today. I think we really do not understand the full dimensions of what is needed for user-friendliness. We have heard much about home computers, and I share the frustrations and experiences of everyone who has pushed a disk in and had the screen read "Disk Read Error," "Error 11," or whatever else can happen. I contend that you can fix those things and still not be user-friendly by a wide margin. I believe that the key to widespread diffusion of the technologies that Walter Baer has been talking about is for us to get to a new level of understanding and appreciation of what user-friendliness really means. Frankly, I think that TV—low enthusiasm but very high acceptance—is a good standard for user-friendliness.

I want to dramatize what I mean by user-friendliness. Let me take the example of the telephone. Think for a minute what happens to you when you want to make a call on the telephone to someone whose number you do not know and whose number is not in that list of 30 numbers you can store in your telephone today. You pull out a big, thick volume that you have tucked away in some obscure spot to keep it from being unsightly, open it up, and run your finger down fine print to try to find the person you want. Then, of course, there is no space to lay the book near the telephone, so you balance it on your knee, hold one finger on the name, pick up the receiver, try to hold it to your ear and dial, and halfway through you drop the telephone book and receiver. I submit that that is not user-friendly.

I believe that we do have to get to a new level of understanding and comprehension of user-friendliness in order to make many home technologies come about. There are a number of technologies on the horizon that can help do that, but the biggest gain will come if we perfect voice input and voice recognition. If we move forward to a state where voice input is low cost and sophisticated, user-friendliness will be just around the corner.

Finally, I agree, by and large, with the assessments that Walter Baer has made. The only one that I would question is whether the impact of learning in the home will be quite as large as he thinks it will be unless we solve the problem of user-friendliness. Moreover, I think that the technologies required to improve user-friendliness are only now emerging and being developed. The software related to such things as expert systems and artificial intelligence—if they come along as dramatically as many of us expect them to—will significantly improve our ability to build truly user-friendly products.

Computers and Business

THEODORE J. GORDON

Computers and automation are so irrevocably entwined with business that it is hard to imagine what business would be like without them. These technologies—computers and automation—are ubiquitous, expected, and necessary, appearing in almost every facet of business enterprise: from recruitment through layoff, from raw material ordering through the manufacturing of products, from identifying sales prospects to order entry and delivery, from competitive analysis to strategic optimization, and from innovation to design applications of computers and automation. They permeate business life, and in doing so have changed it for all time. Yet there is more to come, not only with respect to business applications of computers and automation, but perhaps more importantly with respect to their impact on business itself and on the people who run it.

This is a vast and literally boundless topic, so some structure is necessary if we are to discern even the highlights of prospective change. A three-dimensional space serves as the organizing principle for this paper: Business functions comprise one axis; on another axis lie the technologies; and on the third, the impacts. For convenience, I have divided *business functions* into management, manufacturing, selling, planning, training, and professional support. For *technology* I have defined three major facets: computer hardware and software; programmable automation, which includes robotics; and telecommunications. And in the third dimension, the *impacts* fall on three elements of business: people, those who define and execute the intent of business;

structure, the organization of the business enterprise; and self-concept, the view of what business is about and is capable of attaining.

This structure leads to a set of interesting questions. For example, How will future developments affect people in management? For this discussion I have selected only a few of these intersections, on the basis of their significance, the number of people affected, and the likelihood of the developments in the next 10 to 15 years. This is certainly not a complete set, and it relies heavily on judgments about what is possible and about the complex processes of acceptance and response to new technologies. The changes we have seen to date are staggeringly significant. What is coming, however, not only extends the trends of the past but includes massive changes that, in the aggregate, will define the very nature of business and the relationships between those who serve and are served by it.

My approach will be to set the stage with a discussion of some characteristics of emerging technologies and then to describe some of the changes that seem likely at four intersections in our three-dimensional (business function/technology/impact) space. The intersections and their respective changes are as follows:

- At the intersection of training, computers, and people, the discussion centers on simulation as an aid to training;
- At the intersection of manufacturing, automation, and people, on factory automation and robotics;
- At the intersection of selling, telecommunications, and structure, on retail electronic funds transfer (EFT); and
- At the intersection of management, computer, and self-concept, on modeling in management decision making.

SOME CHARACTERISTICS OF EMERGING TECHNOLOGIES

First, the stage setting. There are three principal hardware trends that characterize electronic hardware today: reduction in cost, improvements in reliability, and increases in packing density (the number of components that can be packed into a given volume). Each of these characteristics of the technology—cost, reliability, and density—has been changing by approximately a factor of 100 each decade since 1960. Studies by The Futures Group indicate that these trends can continue for another two decades or so. As limits are reached during this period, new technologies will offer potential for further breakthroughs. For example, photolithography (the technology required for printing microcircuits on silicon) is limited at present by the distance between lines that can be drawn optically. This, in turn, is fixed by the wavelength of light. Once this limit is reached, conventional photolithography impedes further progress toward miniaturization. However, just behind this conventional technology lies the possibility of using shorter-wavelength energy in these processes; for example, electron beam or X-ray imaging.

In addition to these more or less continuous trends of improved reliability, reduced cost, and increased packing density, it is worth noting two other developments of significance; these are discontinuities that can significantly affect the application of electronics in the future. First, the same techniques that are being used to produce very large scale integrated circuits are also being applied to the manufacture of small mechanical devices. For example, a mass spectrometer, a device for determining the constituent elements of gases and other fluids, has been "printed" on a chip. This is more than simply printing the electronics on a chip, as is commonplace in very large scale integrated circuitry. Rather, the whole machine—valves and all—is part of the printed apparatus.[1] An example of the future use of microsystems in industry is the potential for process instrumentation that floats with a stream of chemicals and telemeters process control data (rather than being point-fixed on the well of a pipe). Another industrial example is temperature and pressure instrumentation built into a grinding wheel or cutting tool to control a feedback system that optimizes metal removal rates, improves precision of manufacture, and extends tool life. In the office these micromechanical devices can be useful in constructing extremely small microphones for dictation, telephones, or security systems (also, micro eavesdropping bugs), or feedback instruments for chairs, printers, and personal, local air conditioning. With this technology, buildings can be instrumented to detect incipient mechanical failures, and quiescent manufactured products can telemeter their state of functioning or readiness in response to an external radio trigger signal. In short, micromechanics not only will permit replication of macromachines on a tiny scale but will stimulate the innovation of entirely new applications that benefit from small size, low price, dispersion, and decentralization.

The second development is the coming of age—probably within the next decade—of artificial intelligence (AI), the simulation of human intelligence by computers. Artificial intelligence requires the ability to sense, operate on sensed information, draw inferences from observations, and perform adaptively in view of these inferences and changing circumstances. AI programs have two general attributes: search and knowledge. Search comes from "defining a space of possibilities large enough to contain the sought-for solution," and "knowledge is necessary to guide the search through the space."[2]

The initial entries in this field are expert systems that capture pragmatic if-then rules of analysis followed by human experts in a given field. Examples of expert systems that are functional or in development abound. Medical diagnostics, personal financial services, geological exploration, legal strategy, and software design are some examples. In these fields and in others yet to come, computers will produce practical and functional answers to real problems better than answers that could be produced by a random sample of professionals.

When expert systems can learn from experience, the decision rules incorporated in the knowledge portion of the program can be much more extensive, and a transition will have occurred from programs that merely emulate the behavior of experts to creative artificial intelligence routines. These routines will arrive at answers better than those that might be created by most human experts. This will not happen tomorrow, but it is reasonable to expect self-learning systems to be in operation within 15 years.

This image of the growth of electronics and its applications and spin-offs depends, of course, on the market. Given demand, the capability of electronics and all of its derivatives grows; without demand, nascent applications wither. Business fosters demand both by offering new products to consumers and by becoming a consumer itself. In either mode business touches and is touched by these technologies and thus is changed, not only in how it does business, but in what it believes business to be. Let us turn now to some intersections in the three-dimensional space discussed earlier.

INTERSECTION: TRAINING, COMPUTERS, PEOPLE

Simulation as an Aid to Training

Take as a starting point high-density TV (in the more distant future, perhaps holographic TV) driven by rapidly accessed videodisks, excellent and sensitive computer simulation programs, and much more effective input-output systems. Put them together and imagine a worker-training system 10 to 15 years hence. The key here is software—superb simulation techniques that permit the creation of accurate environments that stress the student and promote learning. With such software, the emphasis in education switches from teaching to learning. Pilot training serves as a current example. In the future, decreasing equipment costs, better software, and realistic input-output systems mean that applications will be far less monumental and will certainly be applied in business for training of production workers, managers, salespeople, repairmen, and anyone likely to benefit from the stress of practice. Simulation in the context used here means that the user determines the plot of an unfolding story through his or her decisions. This is computer game playing carried to its logical end. In schools you can be with Napoleon or serve as a lieutenant to Washington. In leisure at home, TV becomes active rather than passive—here you can be with Cleopatra or J. R. Ewing. In the workplace you can direct field operations to put out oil well fires, or learn, as a potential manager, how it feels to deal with labor grievances and the chess game of a strike.

Consequences

Personnel training and education will be substantially changed, with competitive advantage falling to the best corporate simulators. Imagine the *new* case study approach in management education: "Have you played Continental Illinois yet?" Grading of personnel in corporations may be according to their successes or failures in simulations. Imagine—in awed tones—"He made Continental survive." Perhaps boredom will become a problem when the game environment makes the adrenaline flow more easily than reality does.

New products and markets will be built around interactive on-the-job experience. When the simulation tools are very good they will also serve as decision-aiding tools. For the repairman not quite sure about which wire to connect, a quick run on the simulator will show the consequences of connecting the wire—a working machine or a blown fuse. The analogous situation for management dilemmas is obvious.

INTERSECTION: MANUFACTURING, AUTOMATION, PEOPLE

Factory Automation and Robotics

When computers were first introduced, there was a great deal of concern that unemployment would result. In fact this was not the case—wherever computers were used, more jobs were created. Common wisdom holds that this situation will always continue, but it might not be so. Factory automation is likely to advance so far beyond current capabilities that the net effect of introducing such new technologies may be to improve total output with less labor required in both a relative and an absolute sense. I will explore the potential for such technology-induced unemployment in this section.

Artificial intelligence will allow machines to perform cognitive

TABLE 1   Forecasts of Robot Technology

| Feature | | State of the Art | |
		1983	2000
Accuracy of manipulation	(Electric)	0.02 inches	0.001 inches
	(Hydraulic)	0.2 inches	0.020 inches
Repeatability of placement	(Electric)	0.005 inches	0.001 inches
	(Hydraulic)	0.050 inches	0.010 inches
Mean time between failures		1,000 hours	5,000 hours
Fault detection and repair		Mostly human	Mostly self-check
Speed for standard pattern		4 seconds	1 second
Programming		External	External and self-taught
Sensing—visual		Silhouette	3-dimensional
Memory capacity and type		Magnetic media	Vastly expanded magnetic and optical media
Information processing		Sequential architecture	Parallel architecture

SOURCE: The Futures Group.

functions; robotics will move from specialized to general-purpose applications. Recent studies at The Futures Group have resulted in projections of robot technology that illustrate the enormous potential for this field. As Table 1 shows, robotic accuracy, repeatability, mean time between failures, time to repair, and speed are expected to increase significantly—more than an order of magnitude in most cases—in the next 15 years.

Within the next decade or so we also can expect to see very simple means for programming robots and, with the advent of artificial intelligence, robots that learn through experience. For example, a robot could be adaptively programmed to change its positioning or sequence in order to minimize rejection rates.

A distinguishing feature of robots is their versatility—their ability to be used in a multiplicity of applications. In the future, robots will become more general-purpose, in the sense that their implements can be utilized in a variety of jobs without much cost penalty. Vision and sensing will improve to make three-dimensional perceptions commonplace.

The number of robots to be employed in the future is not certain by any means, but large increases seem likely. The number of robots in manufacturing quadrupled between 1979 and 1981.[3] Forces encouraging growth include

- Improvements in the technology itself, which increase the number of applications possible with these machines;
- Diminishing costs for given robotic capability as a result of learning-curve improvements;
- Increasing cost of human labor; and
- Growing sophistication on the part of management, facilitating the switch to robotics.

On the negative side, factors that limit the speed of diffusion of this technology include

- The size of the required investment,
- Institutional inertia that slows the adoption of automated technologies, and
- Obsolescence of current equipment.

Now the question is: Will progress in robotics and factory automation in general create jobs or eliminate them? The answer is, of course, it will do both. At constant levels of output it will eliminate jobs, because the robots will perform jobs that human workers currently perform, and automation, properly applied, will improve productivity through increasing output per man-hour. Some people argue that as automation progresses more people will be required to produce the machines and that as people are freed from dull, repetitive, boring, and sometimes dangerous activities, unemployment will not diminish but, in Parkinsonian response, the scope of work and perhaps its quality will increase to occupy the new capacity for work. While this has generally been the case in the past, robotics and the new wave of automation have some new attributes. They promise to be very good, very cheap, and untiring. They can yield manufacturing quality higher than that produced by their human counterparts. Instructing these machines (programming) will be efficient and, through AI, they will be adaptive; thus as creative as the initial programming may be, the machines will progress and learn to do even better. The machines will self-diagnose, learn of impending internal failures through "introspection," and— when necessary—self-repair, to a small extent initially and to a major extent later. More than that, when programmed to do so, they can self-replicate.

Just how many jobs will be displaced by continuing factory automation and robotics? The situation is summarized in Table 2. We have created a scenario with several critical assumptions. The U.S. Department of Labor expects the labor force to grow from its present level of 110 million to about 134 million by the year 2000.[4] We assume that real productivity grows at about 1.5 percent per year as a result

TABLE 2 Employment Scenario: 1982–2000

Year	Labor Force (millions of people)	Gross National Product (billions of 1984 dollars)	Productivity (dollars of GNP/employee, in thousands)	Required Employment (millions of people)	Number of Robots (thousands)	Person Equivalent of Each Robot	Displaced Employment (millions of people)	Net Required Employment (millions of people)	Surplus Labor Force (millions of people)	Percent Unemployed
1982	110	3,059	30.90*	99	5	2	0.01	99	11.01	10.01
1983		3,142	31.36	100						
1984		3,226	31.83	101						
1985	115	3,314	32.31	103	30	2	0.06	102	12.51	10.88
1986		3,403	32.80	104						
1987		3,495	33.29	105						
1988		3,589	33.79	106						
1989		3,686	34.29	107						
1990	122	3,786	34.81	109	100	3	0.30	108	13.54	11.10
1991		3,888	35.33	110						
1992		3,993	35.86	111						
1993		4,101	36.40	113						
1994		4,211	36.94	114						
1995	127	4,325	37.50	115	250	4	1.00	114	12.66	9.97
1996		4,442	38.06	117						
1997		4,562	38.63	118						
1998		4,685	39.21	119						
1999		4,811	39.80	121						
2000	134	4,941	40.40	122	500	5	2.50	120	14.18	10.58

NOTE: Assumptions: GNP growth rate, 2.70 percent; productivity growth, 1.50 percent.
* 1982 productivity figure calculated based on total employment of 99 million people.
SOURCE: The Futures Group.

of automation and that real Gross National Product (GNP) grows at 2.7 percent per year. Furthermore, the number of robots is assumed to grow from a currently installed base of about 5,000 to 500,000 by the turn of the century. We suppose, further, that the effectiveness of each robot also grows; today a robot is equivalent to about two persons, and we have assumed that by the turn of the century a robot can replace five workers. With these assumptions, as Table 2 shows, the contribution of robots is relatively minor. Only about 2 percent of the labor force expected in the year 2000 will be displaced by robotics.

While the picture presented in Table 2 is a homogeneous representation of the labor force as a whole, certain industries will be more affected by robotics than others. In general, these are industries in which mechanization of production yields lower cost, higher quality, diminished production time, improved efficiency, or improved worker safety. In these industries, the impact of robotics on job displacement will be considerable. For example, the production of passenger automobiles has involved a labor force of about 270,000 over the last seven years. On average, this labor force produced about 30 automobiles per employee, while production during this interval varied from 6.2 million to 9.2 million units per year. Now, assume that the number of individuals available for passenger automobile production grows at the same rate as the labor force as a whole. The employees required, however, are affected by level of production of automobiles and by improvements in productivity resulting from factory automation and the introduction of robots. If we assume (1) that production grows at 3 percent a year (so that by the year 2000 more than 10 million units are manufactured in the United States), (2) that productivity grows at 1.5 percent per year (as previously assumed), and (3) that the number of robots used by the industry grows from 5,000 in 1985 to 25,000 in the year 2000, less than half of those who might ordinarily have been assumed to be available for employment in this industry will be required. A major public policy concern, of course, is that those displaced may be the least able to find new employment. Whether this is a barrier to the spread of factory automation depends on many factors, such as the job security provisions of labor contracts in affected industries and the state of the economy. My guess, however, is that robotics and programmable automation will spread rapidly as the return on investment in such systems grows.

Consequences

There is likely to be a growing emphasis on the job security issue and retraining in industries likely to experience displacement. Also, there may be the return to the United States of some jobs previously located outside the country to take advantage of low-cost automated production. Product quality will improve and a new class of employment—such as the blue-collar programmer—will emerge. Factory automation and robotics will constitute a new technological frontier on which the battles for international markets will be fought, since the products these technologies yield will be of lower price, higher quality, and more predictable performance.

INTERSECTION: SELLING, TELECOMMUNICATIONS, STRUCTURE

Retail Electronic Funds Transfer (EFT)

There is an unusual and, to some extent, unexpected confluence of technological and consumer trends that may affect the retailing environment in the immediate future. I believe that automated debit purchasing may come on the scene faster than many people expect. Here are some of the factors that lead to this position:

- Automatic teller machines (ATMs) have spread (47,000 units in place in 1983) and have become much more widely accepted by consumers.[5]
- The principal use of ATMs has been to withdraw cash from personal accounts.
- Many retail establishments are considering or are installing ATMs on their premises—small bank branches—in order to provide a means for their customers to obtain cash in their stores.
- The communication networks required to support the ATMs and the software necessary to properly debit accounts exist and are proven.

It takes only a small step of imagination to move the ATM *into* the cash register so that at time of purchase consumers can simply insert the ATM card into the proper slot, punch in their personal identification number, and be charged for the purchase directly, just as if they had gone to the ATM and withdrawn cash. In this view, there will be no such thing as a special debit card. The bank-issued ATM card takes its place and the era of electronic fund transfers at the retail level will have come on the scene smoothly, with minimum fanfare, and with relatively high consumer acceptance. More than just a convenience to shoppers, this is a crucial and catalytic step to a society that, while not cashless, certainly functions with much less cash.

An important by-product of this development for business will be the availability of really exquisite information about who buys what, where, and when—the basis for a potential revolution in marketing research. Where will this trend surface first? Probably in supermarkets and gas stations. It then spreads in the retail environment wherever population density and purchase traffic are high enough to warrant the investment. Then it may spread to vending machines of this new era. These vending machines will be operated by coins as well as by ATM cards. What better way could be found to reduce store theft? The image for these vending machines will be upscale, and I would guess that these machines will typically carry much higher-value items than do machines of today. One feature of these machines that will make them unique is their ability to be easily programmed to offer a wide variety of merchandise (sweaters, socks, small electric appliances, stationery). In this way the machines can be generic, and the price can be set for whatever merchandise the retailer wants to sell.

Consequences

The rate of growth in the number of checks written and processed will be greatly reduced, as will the amount of mail. There will be a major new role for secure telecommunications networks, and today's credit cards will be transformed into their debit-credit equivalent. Once the networks have linked consumers with their accounts at points of sale, whole markets can go electronic, allowing buyers and sellers to meet electronically, and bid and auction until deals are made remotely. It seems to me that this kind of market, based on network connections, could be very well suited to real estate, tax shelters, and any other high-value transaction in which "feeling the merchandise" is not essential.

INTERSECTION: MANAGEMENT, COMPUTER, SELF-CONCEPT

Modeling in Management Decision Making

The goals of a business are, by and large, determined by what management believes is feasible at acceptable levels of investment and risk. These perceptions are, in turn, informed by data available to management about their customers, about competing business, and about the environment in which they operate. The sensory capacity of the business to determine what is happening around it has improved and will continue to improve. Beyond this, the ability to distill nuggets of pertinent information out of such data enhances not only decision-making capacity but the goals on which such decisions are based. The secret here is not knowledge about customers, competitors, and the environment—real knowledge about such matters is restricted to the past—but rather the quantification of uncertainty and risk. New analytic techniques will facilitate the introduction of uncertainty into decision processes, and from the beginnings that are already in place today, corporate actions will be weighed not only on the basis of expected return but on levels of acceptable risk.

Here is an example. Suppose that a forecast of demand for an existing product has just been made. Using probabilistic tools, the forecast recognizes irresolvable uncertainties: the potential entry of competitors, the emergence of a new technology that could overtake the product, the potential for a fad that could spark unprecedented demand. Such factors produce two scenarios for the future. Suppose further that the first scenario requires building a new plant and the second does not. Has quantification of the level of uncertainty helped make the decision about the plant? Of course. One could reason as follows. Case 1: I believe the first scenario and build a plant, but the second scenario occurs. Case 2: I believe the second scenario and do not build the plant, but the first scenario occurs. Clearly one situation is better than the other, and even in this simple example of risk analysis, quantification of uncertainty helped resolve the issue. For risk analysis to become very accurate and helpful, models must improve and be trusted by managers, and data about environmental factors, competition, and customers must be collected regularly. All of these developments are happening and will accelerate.

Consequences

The use of corporate intelligence gathering, not as espionage but as a routine and accepted business function, will increase. Decision making will become explicitly risk-conscious, and decisions will be evaluated not in terms of return on investment (ROI) but in terms of ROI probability distributions. Intuition and the "gut call" will remain, of course, but in a probabilistic context—some high-risk opportunities will be seen as worth taking, while some lower-risk options will be judged not worth it. For better or for worse, probabilistic methods will make decision making more explicit and management more self-conscious and more accountable for its performance.

CONCLUSION

This has inevitably been a very rapid tour of some of the more important intersections in the function—technology—impact space, but several more cells deserve some mention.

- *Computer-Aided Design and Computer-Aided Manufacture (CAD/CAM).* The design shop and the shop floor are being modified in function and form to capture the advantages of efficiency, quality, and accuracy afforded by the new technology. CAD/CAM went from a $1-million industry in 1973 to $1.2 billion in 1983. It is expected to grow by a factor of 10 between 1985 and 1995.
- *Electronic Mail.* Imagine an electronic typewriter that is likely to be on the market in 5 years or so. It has a matrix printer, memory for several lines, built-in word processing, built-in spelling correction, and several other easy-to-accomplish software features. This typewriter is sold not as an office word processor or computer peripheral, but simply as a consumer-oriented portable. Its price is less than $100, and it is a standard gift to high school graduates about to leave for college. It is not difficult to imagine the sale of several million of these devices, perhaps several scores of millions, within the next 10 years. With a modem chip these machines can be plugged into standard telephone jacks. Electronic mail, for better or worse, will have arrived overnight.
- *Group Decision Making.* Automated voting machines currently exist (e.g., the CONSENSOR) that permit participants in a meeting to provide judgments in response to questions posed by a monitor. In some of these machines, an individual's input can be weighted on the basis of expertise or knowledge. With improved expert systems and artificial intelligence, group interactions can be computer-augmented. Individual weights can be set by experience or testing, and the group itself can be integrated with expert systems and judgments drawn particularly against the profile of issues being addressed. Also, of course, on-line data can be called up if necessary to provide background for the group as a whole. These techniques tend to subdue the normal psychological problems that accompany group interactions and to promote smoother, more precise, and probably more accurate decision making. The consequence will be that teams, even teams composed of individuals located at remote places, become more important, at the expense of the individual.

The way business does its work is being profoundly and permanently changed. On the factory floor it is being changed by numerically controlled machines, analog and digital sensing devices, automated testing gear, design systems, material-handling systems, inventory control systems, and automated stockrooms. In the office changes are being driven by electronic filing, automated scheduling of meetings, direct access to information, word processing, and, soon, idea processing. The computer affects almost every job and every worker. The nature of the jobs, where they are done, the way they are accomplished, the expectations of job supervision, and the skills required to perform the tasks are all changing. Additionally, the information available for doing tasks and the precision and timeliness with which they can be done are changing.

Granted business processes are changing, but is business itself changing? After all, business takes raw materials, adds value, and sells products. At this level, is anything likely to be different? Early critics of computers—computers using cards that warned against folding, mutilating, or spindling—were concerned about the way that computers would regiment and standardize us all, force us into providing rigid inputs that computers could understand. Now it is clear that computers provide the ability to manipulate and track information, and a variety of individual needs can be easily accommodated. Computers do not standardize; they promote diversity. We are coming closer to the time when the *user* can design the product and when, in sensitive and high-quality work environments, the worker can change his or her environment and utilize an array of information that makes that person's contribution unique. Through the computer, business gains diversity.

Business may also gain responsibilities. The prescription to take raw materials, add value, and sell products may be too simple for a future age. For all of its elegance, accomplishments, and promise, the computer can cause human obsolescence, displacement of workers, and unemployment. The role that business will have in accommodating these discontinuities is ideological as well as economic, and it is far from clear how that role will evolve. Somewhere in this chaotic, complex, and uncertain mix lies business of the future.

NOTES

1. J. B. Angell, S. C. Terry, and P. W. Barth. 1983. Silicon micromechanical devices. *Scientific American,* April:44-55.
2. Committee on Science, Engineering, and Public Policy; National Academy of Sciences; National Academy of Engineering, and Institute of Medicine. Cognitive science and artificial intelligence. In *Research Briefings 1983*. Washington, D.C.: National Academy Press, p. 25.
3. S. A. Levitan, and C. M. Johnson. 1982. Future of work: does it belong to us or to the robots. *Monthly Labor Review,* Vol. 105:10-14.
4. H. N. Fullerton. 1980. The 1995 labor force: a first look. *Monthly Labor Review,* Vol. 103, No. 12:11-21.
5. R. M. Garsson. 1983. Fast growing ATMs are now as ubiquitous as Xerox machines—Georgia firm aims new computer at bank calling officers. *American Banker,* December 21, 1983:8ff.

Comments

RUTH M. DAVIS
President
The Pymatuning Group

In spite of the scope of the papers presented here I still do not have the big picture of how information technology is going to affect us. In fact, I believe that I am fortunate in not understanding it; I may embody the discontinuity that is going to allow technology and people to get along together. Like everyone else, I need more time to adapt than advancing technology normally permits. Admission of the need to gradually make the necessary moves to adapt and to learn how to use this technology is going to be the key to its success.

Ted Gordon has offered some superb insights into one of the arenas where this technology is making the greatest change. It is interesting, though, that behind this facade of change there are some very subtle inferences. For example, as we have heard, it is very easy for managers to believe they are making changes in a company without actually doing anything. They manipulate the information, talk about it, tell people, simulate it, and generate interaction; they do this without making a single change in the company. We can all generate a tremendous number of information activities without any real activity in the marketplace. The result can be a different version of "much ado about nothing"—that is, "much ado *with* nothing." I think we have to be terribly careful that we do not jump from what we know how to do—to work where we can see results in real time—to manipulating results faster than real time in terms of people's abilities to react.

One of the most interesting points that emerges from the combination of Walter Baer's and Ted Gordon's papers is the blurring of the differentiation, which I have always been uncomfortable with, between work and leisure. That differentiation is rapidly disappearing. Many times I would rather sit at a computer at home and play with a simulator than sit on the patio and smack mosquitoes. There will be many instances when the kind of work that we did in the past is going to be replaced by fun because of the manner in which the result is attained. If we cannot differentiate between leisure and working, however, we will really have to adjust to a tremendous change in the near future. From the anticipated confusion will evolve the real world of tomorrow. I think the papers in this volume give you all the background that you need to determine your role in the mixed marketplace and the mixed business/ leisure world of the future.

Much has been said here about information technology and business; Ted Gordon has helped focus this picture. I commend to you his concept that will lead you through the spectrum of technology in business, from the board of directors on one side of the marketplace to the consumer on the other side. You can use it in your own business to see where lies the real power for change. It has been difficult to bring technology to play in the boardroom and to determine what companies should do to improve their market position and products. It is also very difficult to simulate a marketplace and the effects of policy decisions. I can tell when I have crashed my simulated airplane into a simulated Sears Tower, but I cannot tell when I am making a bad prediction in the boardroom.

It is also important to consider young consumers, those individuals on the other side of the marketplace from the boardroom. We really cannot predict their behavior even though they are already consumers of information technology. At three or four years of age they use their terminals and watch video, cutting out commercials and listening only to what they want. As those young consumers, who are now making the process of learning and buying continuous, grow older, they are going to be utilizing information technology to lead them to many decisions immediately, and they are going to grab hold of the marketplace and swing it around like a child swinging a lion by the tail. For a long time we have had a more homogeneous consumer group that the selling side of the marketplace could manipulate: a group that ranged from the age of 15 to the age of 60. The information technology that you have heard about is the technology of the individual, and it will result in dramatic changes in the consumer marketplace that will, in turn, force changes in business.

About the Authors

WALTER S. BAER is director of advanced technology at the Times Mirror Company, Los Angeles, California. Prior to joining Times Mirror in 1981, he was director of the Energy Policy Program at the Rand Corporation in Santa Monica, California. He has published widely in the fields of energy, telecommunications, and science and technology policy. His book *Cable Television: A Handbook for Decisionmaking* received a 1975 Preceptor Award from the Broadcast Industry Conference. Before joining Rand Dr. Baer served on the staff of the Office of Science and Technology in the Executive Office of the President and as a White House Fellow with Vice-President Hubert Humphrey. Dr. Baer holds a B.S. degree from the California Institute of Technology and a Ph.D. degree in physics from the University of Wisconsin. He is currently a member of the faculty of the Rand Graduate Institute.

ANNE WELLS BRANSCOMB is a lawyer specializing in communications law, a fellow of the Gannett Center for the Study of the Media at Columbia University, and a recent chairman of the Communications Law Division of the American Bar Association Science and Technology Section. Mrs. Branscomb serves on the Steering Committee of the Annenberg Scholars Program of the Annenberg School of Communications at USC and on the Advisory Committee of the Communications Law Program of UCLA. She is a contributing editor of *The Information Society* and the *Journal of Communication*, and a trustee of the Rensselaer Polytechnic Institute. She is currently writing a book, *Teletribes and Telecommunities*, on the social and political impact of communications technologies and is editor of *Toward a Law of Global*

Communications Networks. She is a member of the Commission on Freedom and Equality of Access to Information. Mrs. Branscomb is an honor graduate of the George Washington University Law School and holds degrees in political science from Harvard University and the University of North Carolina.

HARLAN CLEVELAND is dean of the University of Minnesota's Hubert H. Humphrey Institute of Public Affairs, and professor of public affairs; he has been at Minnesota since August 1980. A Princeton University graduate, he was a Rhodes Scholar at Oxford University in the late 1930s; an economic warfare specialist in Washington, D.C., and United Nations Relief Administrator in Italy and China in the 1940s; and a foreign aid manager, magazine editor and publisher, and dean of the Maxwell School of Citizenship and Public Affairs at Syracuse University in the 1950s. Mr. Cleveland served as Assistant Secretary of State for International Organization Affairs in the administration of President John F. Kennedy, and as U.S. ambassador to NATO under President Lyndon B. Johnson. From 1969 to 1974 he was the president of the University of Hawaii, and from 1974 to 1980 he was director of the Program in International Affairs of the Aspen Institute for Humanistic Studies. He is author or editor of 13 books. His latest book, from which his paper for this volume is derived, is *The Knowledge Executive: Leadership in an Information Society* (New York: E. P. Dutton, 1985).

THEODORE J. GORDON is president of The Futures Group, which he founded in 1971. The Futures Group is a management consulting firm in the fields of planning, futures research, policy analysis, and project implementation. He has been associated with futures research and policy analysis for many years, and has made substantive and methodological contributions to both fields. He is one of the innovators or co-innovators of several methods of forecasting, including cross-impact analysis, trend impact analysis, and probabilistic system dynamics. Mr. Gordon helped establish the Institute for the Future, where he served as vice-president and senior research fellow prior to 1971. Before joining the institute, Mr. Gordon directed engineering programs at the McDonnell-Douglas Astronautics Company, serving variously over 16 years as chief engineer of the Saturn Program, test conductor for the THOR and THOR-Launch Systems, and director of Advanced Space Systems and Launch Vehicles. Mr. Gordon earned a B.S. in aerodynamics from Louisiana State University and an M.S. in aerodynamics from the Georgia Institute of Technology.

MELVIN KRANZBERG is the Callaway Professor of the History of Technology at Georgia Institute of Technology. He is the principal founder of the Society for the History of Technology; edited its quarterly journal, *Technology and Culture,* from 1959 to 1981; and became president of the society in 1983. Dr. Kranzberg was one of the original members of the History Advisory Committee of the National Aeronautics and Space Administration, serving as its chairman from 1966 to 1969; in 1984 he was reappointed chairman of that committee and made a member of the NASA Advisory Council. In 1979–1980 Dr. Kranzberg was national president of Sigma Xi, the Honorary Scientific Research Society. Professor Kranzberg received his A.B. from Amherst College, and his M.A. and Ph.D. from Harvard University.

JOHN S. MAYO is executive vice-president, Network Systems at AT&T Bell Laboratories in Murray Hill, New Jersey. Since joining Bell Laboratories in 1955, Dr. Mayo has been director of the Oceans Systems Laboratory, executive director of the Ocean Systems Division, executive director of the Toll Electronic Switching Division, and vice-president of Electronics Technology. He assumed his present position in May 1979. Dr. Mayo received his B.S., M.S., and Ph.D. degrees in electrical engineering from North Carolina State University. He was elected to the National Academy of Engineering in 1979.